T0348900

CARBON-BASED MATERIALS
FOR MICROELECTRONICS

ELSEVIER SCIENCE Ltd
The Boulevard, Langford Lane
Kidlington, Oxford OX5 1GB, UK

ISBN 0-08-043614-5
Reprinted from: CARBON, vol. 37/5

Library of Congress Cataloging in Publication Data
A catalog record from the Library of Congress has been applied for.

British Library Cataloguing in Publication Data
A catalogue record from the British Library has been applied for.

⊗ The paper used in this publication meets the requirements of ANSI/NISO Z39.48-1992 (Permanence of Paper).
Transferred to digital printing 2006
Printed and bound by CPI Antony Rowe, Eastbourne

CARBON-BASED MATERIALS FOR MICROELECTRONICS

PROCEEDINGS OF SYMPOSIUM K ON
CARBON-BASED MATERIALS FOR MICROELECTRONICS
OF THE E-MRS 1998 SPRING CONFERENCE

STRASBOURG, FRANCE, JUNE 16-19, 1998

Edited by

J. ROBERTSON
University of Cambridge, United Kingdom

J. FINK
IFW Dresden, Germany

E. KOHN
Universität Ulm, Germany

1999

ELSEVIER
AMSTERDAM - LAUSANNE - NEW YORK - OXFORD - SHANNON - SINGAPORE - TOKYO

Vol. 50: High Temperature Electronics (eds. K. Fricke and V. Krozer)

Vol. 51: Porous Silicon and Related Materials (eds. R. Hérino and W. Lang)

Vol. 52: Small Scale Structures (eds. N.F. de Rooij, J.-M. Moret, H. Schmidt, W. Göpel, A.L. Greer, K. Samwer and C.V. Thompson)

Vol. 53: Ion Beam Processing of Materials and Deposition Processes of Protective Coatings (eds. P.L.F. Hemment, J. Gyulai, R.B. Simonton, I. Yamada, J.-P. Thomas, P. Thévenard, W.L. Brown, P.B. Barna, Y. Pauleau and G. Wahl)

Vol. 54: Selected Topics in Group IV and II-VI Semiconductors (eds. E. Kasper, E.H.C. Parker, R. Triboulet, P. Rudolph and G. Müller-Vogt)

Vol. 55: Laser Ablation (eds. E. Fogarassy, D. Geohegan and M. Stuke)

Vol. 56: C, H, N and O in Si and Characterization and Simulation of Materials and Processes (eds. A. Borghesi, U.M. Gösele, J. Vanhellemont, A.M. Gué and M. Djafari-Rouhani)

Vol. 57: Porous Silicon: Material, Technology and Devices (eds. R. Hérino, W. Lang and H. Münder)

Vol. 58: Organic Materials and Fullerenes (eds. D. Bloor, G. Leising, G. Zerbi, C. Taliani, P. Bernier and R. Taylor)

Vol. 59: Frontiers in Electronics: High Temperature and Large Area Applications (eds. J. Camassel, K. Fricke, V. Krozer, J.L. Robert, B. Drévillon, B. Equer, I. French and T. Kallfass)

Vol. 60: Semiconductors and Organic Materials for Optoelectronic Applications (eds. R.L. Aulombard, B.C. Cavenett, B. Gil, R. Triboulet, G. Leising and F. Stelzer)

Vol. 61: Group IV Heterostructures, Physics and Devices (Si, Ge, C, α-Sn) (eds. J.-M. Lourtioz, G. Abstreiter and B. Meyerson)

Vol. 62: Magnetic Ultra Thin Films, Multilayers and Surfaces (eds. M.A.M. Gijs and F. Petroff)

Vol. 63: High Temperature Superconductor Thin Films: Growth Mechanisms-Interfaces-Multilayers (eds. H.-U. Habermeier and M.L. Hitchman)

Vol. 64: Laser Processing of Surfaces and Thin Films (eds. C.N. Afonso, E. Matthias and T. Szörényi)

Vol. 65: New Trends in Ion Beam Processing of Materials and Beam Induced Nanometric Phenomena (eds. F. Priolo, J.K.N. Lindner, A. Nylandsted Larsen, J.M. Poate, E.E.B. Campbell, R. Kelly, G. Marletta and M. Toulemonde)

Vol. 66: Advanced Materials for Interconnections (eds. Th. Gessner, J. Torres and G. Crean)

Vol. 67: New Developments in Porous Silicon. Relation with other Nanostructured Porous Materials (eds. L.T. Canham and D. Bellet)

Vol. 68: Fullerenes and Carbon Based Materials (eds. P. Delhaes and H. Kuzmany)

Vol. 69: Recent Developments in Thin Film Research: Epitaxial Growth and Nanostructures, Electron Microscopy and X-Ray Diffraction (eds. G. Ritter, C. Matthai, O. Takai, A.M. Rocher, A.G. Cullis, S. Ranganathan and K. Kuroda)

Vol. 70: Computational Modeling of Issues in Materials Science (eds. H. Dreyssé, Y. Kawazoe, L.T. Wille and C. Demangeat)

Vol. 71: Light-Weight Materials for Transportation and Batteries and Fuel Cells for Electric Vehicles (eds. R. Ciach, H. Wallentowitz, T. Hartkopf, A. Moretti, J.G. Wurm and M. Wakihara)

Vol. 72: Biomaterials and Biodegradable Polymers and Macromolecules: Current Research and Industrial Applications (eds. F. Burny, D. Muster and A. Steinbüchel)

Vol. 73: Coatings and Surface Modifications for Surface Protection and Tribological Applications (eds. J.P. Rivière, L. Pranevicius, J.-M. Martínez-Duart and A. Grill)

Vol. 74: III-V Nitrides Semiconductors and Ceramics: from Material Growth to Device Applications (ed. B.K. Meyer)

Vol. 75: Materials, Physics and Devices for Molecular Electronics and Photonics (eds. J. Zyss and F. Garnier)

Vol. 76: Defects in Silicon: Hydrogen (eds. J. Weber and A. Mesli)

Vol. 77: Light Emission from Silicon: Progress towards Si-Based Optoelectronics (eds. J. Linnros, F. Priolo and L. Canham)

Vol. 78: Growth, Characterisation and Applications of Bulk II–VIs (eds. R. Triboulet, P. Capper and G. Müller-Vogt)

Vol. 79: Thin Films Epitaxial Growth and Nanostructures (eds. E. Kasper, K.L. Wang and H. Hasegawa)

Vol. 80: Thin Film Materials for Large Area Electronics (eds. B. Equer, B. Drevillon, I. French and T. Kallfass)

Vol. 81: Techniques and Challenges for 300 mm Silicon: Processing, Characterization, Modelling and Equipment (eds. H. Richter, P. Wagner and G. Ritter)

Vol. 82: Surface Processing: Laser, Lamp, Plasma (eds. I.W. Boyd, J. Perrière and M. Stuke)

Vol. 83: Materials Aspects in Microsystem Technologies (eds. D. Barbier, W. Lang, J.R. Morante, P. Temple-Boyer and G. Mueller)

Vol. 84: Rapid Thermal Processing (eds. A. Slaoui, R.K. Singh, T. Theiler and J.C. Muller)

Vol. 85: Ion Implantation into Semiconductors, Oxides and Ceramics (eds. J.K.N. Lindner, B. Svensson, P.L.F. Hemment and H.A. Atwater)

Vol. 86: Carbon-Based Materials for Microelectronics (eds. J. Robertson, J. Fink and E. Kohn)

Vol. 87: Nitrides and Related Wide Band Gap Materials (eds. A. Hangleiter, J.-Y. Duboz, K. Kishino and F.A. Ponce)

Vol. 88: Molecular Photonics for Optical Telecommunications: Materials, Physics and Device Technology (eds. F. Garnier and J. Zyss)

Vol. 89: Materials and Processes for Submicron Technologies (ed. J.M. Martínez-Duart)

CARBON

VOLUME 37, NUMBER 5   1999

Contents

Papers presented at the European Materials Research Society 1998 Meeting, Symposium K: Carbon-based Materials for Microelectronics

J. Robertson, J. Fink E. Kohn and D. Walton	715	Preface
C. Schlebusch, J. Morenzin, B. Kessler and W. Eberhardt	717	Organic photoconductors with C_{60} for xerography
H. Giefers, F. Nessel, S.I. Györy, M. Strecker, G. Wortmann, Y.S. Grushko, E.G. Alekseev and V.S. Kozlov	721	Gd–L_{III} EXAFS study of structural and dynamic properties of Gd@C_{82} between 10 and 300 K
G. Costantini, S. Rusponi, E. Giudice, C. Boragno and U. Valbusa	727	C_{60} thin films on Ag(001): an STM study
M. Knupfer, T. Pichler, M.S. Golden, J. Fink, A. Rinzler and R.E. Smalley	733	Electron energy-loss spectroscopy studies of single wall carbon nanotubes
L.P. Biró, G.I. Márk, J. Gyulai, N. Rozlosnik, J. Kürti, B. Szabó, L. Frey and H. Ryssel	739	Scanning probe method investigation of carbon nanotubes produced by high energy ion irradiation of graphite
O.M. Küttel, O. Gröning, Ch. Emmenegger, L. Nilsson, E. Maillard, L. Diederich and L. Schlapbach	745	Field emission from diamond, diamond-like and nanostructured carbon films
U. Hoffmann, A. Weber, C.-P. Klages and T. Matthée	753	Field emission of nitrogenated amorphous carbon films
J. Robertson	759	Electron field emission from diamond and diamond-like carbon for field emission displays
S. Logothetidis, M. Gioti, P. Patsalas and C. Charitidis	765	Insights on the deposition mechanism of sputtered amorphous carbon films
T. Heitz, B. Drévillon, C. Godet and J.E. Bourée	771	C–H bonding of polymer-like hydrogenated amorphous carbon films investigated by in-situ infrared ellipsometry
R.U.A. Khan, A.P. Burden, S.R.P. Silva, J.M. Shannon and B.J. Sealy	777	A study of the effects of nitrogen incorporation and annealing on the properties of hydrogenated amorphous carbon films
R. Kalish	781	Doping of diamond
M. Kunze, A. Vescan, G. Dollinger, A. Bergmaier and E. Kohn	787	δ-Doping in diamond
L. Ley, R. Graupner, J.B. Cui and J. Ristein	793	Electronic properties of single crystalline diamond surfaces

H.J. Looi, L.Y.S. Pang, A.B. Molloy, F. Jones, M.D. Whitfield, J.S. Foord and R.B. Jackman   801   Mechanisms of surface conductivity in thin film diamond: Application to high performance devices

J.-P. Lagrange, A. Deneuville and E. Gheeraert   807   A large range of boron doping with low compensation ratio for homoepitaxial diamond films

S. Salvatori, M.C. Rossi and F. Galluzzi   811   Minority-carrier transport parameters in CVD diamond

R.B. Jackman, H.J. Looi, L.Y.S. Pang, M.D. Whitfield and J.S. Foord   817   High-performance devices from surface-conducting thin-film diamond

S. Waidmann, K. Bartsch, I. Endler, F. Fontaine, B. Arnold, M. Knupfer, A. Leonhardt and J. Fink   823   Electron energy-loss spectroscopy in transmission of undoped and doped diamond films

A. Ilie, J. Robertson, N. Conway, B. Kleinsorge and W.I. Milne   829   Photoconductivity and recombination in diamond-like carbon

Yu.I. Prilutski, E.V. Buzaneva, L.A. Bulavin and P. Scharff   835   Structure, dynamics and optical properties of fullerenes C_{60}, C_{70}

C.W. Chen and J. Robertson   839   Doping mechanism in tetrahedral amorphous carbon

P. Gantenbein, S. Brunold, U. Frei, J. Geng, A. Schüler and P. Oelhafen   843   Chromium in amorphous hydrogenated carbon based thin films prepared in a PACVD process

S.F. Kharlapenko and S.V. Rotkin   847   Frenkel-excitations of C_N ($N=12,60$) clusters

H. Lange, P. Byszewski, E. Kowalska, J. Radomska, A. Huczko and Z. Kucharski   851   Evaluation of various processes for $C_{60}Fe$ production

C.L. Xu, B.Q. Wei, R.Z. Ma, J. Liang, X.K. Ma and D.H. Wu   855   Fabrication of aluminum–carbon nanotube composites and their electrical properties

M. Kerford and R.P. Webb   859   An investigation of the thermal profiles induced by energetic carbon molecules on a graphite surface

P. Patsalas, S. Logothetidis, P. Douka, M. Gioti, G. Stergioudis, Ph. Komninou, G. Nouet and Th. Karakostas   865   Polycrystalline diamond formation by post-growth ion bombardment of sputter-deposited amorphous carbon films

E. Evangelou, N. Konofaos, S. Logothetidis and M. Gioti   871   Electrical behaviour of metal/a-C/Si and metal/CN/Si devices

E. Kowalska, Z. Kucharski and P. Byszewski   877   Comparison of fullerene–iron complexes modeling with experimental results

Pergamon

Carbon 37 (1999) 715

CARBON

Preface

Carbon-based Materials for Microelectronics

There have been great advances in our understanding and use of inorganic carbon in recent years, following the development of the vapour synthesis of diamond, the discovery of the C_{60} molecule and the discovery of carbon nanotubes.

This issue contains the papers from the Symposium K 'Carbon-based Materials for Microelectronics' of the European Materials Research Society meeting which was held on 16–19 June 1998, Strasbourg, France.

The symposium covered fullerenes, nanotubes, diamond and amorphous carbon. It was able to show the similarities between the sp2 and sp3 forms of carbon, and between the crystalline, nano-structured and amorphous forms. Carbon is unique in having such a range of covalently bonded forms.

The symposium consisted of 34 oral papers, of which 10 were invited, and 35 poster papers. The papers in this proceedings cover many of the recent developments in carbon, for example the effect of doping on the electronic structure of nanotubes, the discovery of phosphorus doping of diamond, the surface structure and electronic structure of diamond, and the field emission properties of diamond and diamond-like carbon.

The applications of carbon lag some way behind those of other materials, but the symposium highlighted the uses or potential of carbon in xerography, in field emission displays and in photoconductivity-based sensors and radiation detectors.

J Robertson, Cambridge
J Fink, Dresden
E Kohn, Ulm
D Walton, Sussex

Pergamon

Carbon 37 (1999) 717–720

CARBON

Organic photoconductors with C_{60} for xerography

C. Schlebusch*, J. Morenzin, B. Kessler, W. Eberhardt

Forschungszentrum Jülich, Institut für Festkörperforschung, D-52425 Jülich, Germany

Received 16 June 1998; accepted 3 October 1998

Abstract

Our aim is to improve the charge-generation efficiency of organic-photoreceptor materials used in xerography and for laser printers by mixing the material with buckminsterfullerene C_{60}. The improvement is based on an electron transfer from the excited excitonic state of the photoreceptor molecule to the C_{60} thus reducing the recombination probability. Typical materials studied are various metal-phthalocyanines (M-Pc, with M=TiO, VO, Ni, Fe, Cu) and metal-free H_2-Pc. Using techniques like X-ray absorption near edge spectroscopy (XANES) and photoelectron spectroscopy with ultraviolet radiation (UPS) we demonstrate that the electron transfer is energetically possible and no ground state electron transfer occurs. The observation of fluorescence quenching upon doping with C_{60} validates the electron transfer-model. © 1999 Elsevier Science Ltd. All rights reserved.

Keywords: A. Fullerene; C_{60}; C. Electron spectroscopy

1. Introduction

Charge generation upon light excitation is the basic step for the performance of photoreceptors for xerography or laser printers. The use of organic materials, like phthalocyanines (Pc), as the active component is favoured compared to inorganic materials because they are less toxic and their sensitivity reaches into the infrared region, which enables the use of inexpensive infrared lasers [1]. In organic materials the incoming light excites excitonic states that may recombine instead of producing free charge carriers. By mixing the strong electron acceptor C_{60} with the material we hope to decrease the recombination rate due to a fast electron transfer onto the fullerene. Such an electron transfer has already been observed for some organic photoconducting polymers [2–5]. C_{60} is a very suitable electron acceptor for this purpose since it has a high electron affinity and a low chemical reactivity, which is important since a chemical reaction might induce a ground-state electron transfer that enhances an unwanted dark conductivity. To facilitate this transfer process the charge separated state with the electron on the C_{60} and the hole on the Pc has to be lower in energy than the intermediate excitonic state (see Fig. 1). With our spectroscopic techniques of X-ray absorption near edge spec-

troscopy (XANES) and ultraviolet photoelectron spectroscopy (UPS) we are able to test whether this condition for an improvement of a specific material by an addition of C_{60} is fulfilled. Information about the energy of the intermediate excitonic state is obtained by optical transmission spectra. The electron transfer is proven by the detection of fluorescence quenching as a function of the amount of C_{60} that is mixed into the material. The

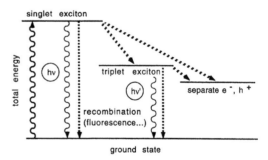

Fig. 1. Total energy scheme of photoexcited and charge separated states (electron on the C_{60}, hole on the Pc-molecule). An electron transfer onto the C_{60} only occurs when the charge separated state is lower in energy than the intermediate state. A fast electron transfer may lower the recombination rate (via fluorescence and non-radiation processes) and therefore enhance the probability of generating free charge carriers.

*Corresponding author. Tel: +49 2461 614247; fax: +49 2461 612620.

improved charge generation efficiency can be demonstrated by measuring the photoconductivity.

2. Experimental

Samples were either prepared by drop casting or spin-coating from commercial dispersions [6] that contain crystallites of τ-H_2-Pc (also known as X-phase) and a polymeric binder [6] or by vacuum sublimation of the pure M-Pc or H_2-Pc onto metal substrates. A part of the dispersion samples was mixed with 5% (by weight) of C_{60}. A homogeneous mixture with higher amounts of C_{60} for a measurement of the fluorescence quenching was achieved by a co-evaporation of Pc and C_{60}. The amount of C_{60} contained in the samples was estimated from the evaporation rate that is measured by a quartz microbalance. This sample is 100 nm thick on a quartz substrate. For the UPS-measurements a low coverage of C_{60} (0.5–2 monolayers) was vacuum sublimed onto the surface of the specific Pc. By sublimation a mixture of the α- and the β-phase of H_2-Pc was obtained with the α-phase being dominant. Since the metastable τ-phase did not survive sublimation, a sample without the binder was prepared by drop casting for the UPS measurement.

X-ray absorption near edge spectra (XANES) were measured using the tuneable synchrotron radiation from the HE-PGM 3 monochromator at the Berlin electron storage ring for synchrotron radiation (BESSY). In XANES a core-level electron was excited to the unoccupied states. Therefore the absorption-cross section depends on the density of the unoccupied states and can be measured via the electron yield as a function of the photon energy. The energy resolution for the spectra shown here is about 0.3 eV.

The photoelectron spectra (UPS) were measured using a He-resonance light source ($h\nu = 21.2$ eV) and a hemispherical electron analyser. All measured kinetic energies were converted to a binding-energy scale with the Fermi-level as a reference. The energy resolution is 0.15 eV.

Electron spectroscopy is a surface sensitive method. The information depth depends on the kinetic energy of the electrons due to their escape length from the material. Therefore with UPS the information is restricted to only a few layers close to the surface, which makes preparation and handling of the samples under ultra-high vacuum (UHV) conditions preferable. The total yield of electrons which is used for the XANES signal, on the other hand, is dominated by electrons with very low kinetic energies of only a few eV that transport more bulk information due to their larger escape depth. This makes XANES a more suitable method for the analysis of the dispersion samples which have been handled in air before insertion into the UHV system for analysis.

An Ar-ion laser that produces photons at 514 nm with 500 mW was used for the fluorescence spectra. The fluorescence signal was measured with a Jobin-Ivon monochromator and a liquid-nitrogen cooled CCD-camera. A correction for the spectral response of the setup was made using a Tungsten-halogen light source as a reference. Due to slightly different focusing conditions between the reference measurement and the measurement with the Pc-samples an artificial structure was created in the spectra, centred at about 920 nm.

3. Results and discussion

In Fig. 2 a XANES spectrum from a dispersion sample containing H_2-Pc, binder and 5% C_{60} is displayed (solid line). This measured spectrum can be almost perfectly reproduced by a mathematical superposition (broken line) of the XANES spectra of the single components shown in the inset. Since no extra features and no peak broadening is obtained when the three materials are in contact with each other we conclude that no chemical reaction occurs. Therefore we have no indication for a ground-state charge transfer and we do not expect an enhanced dark conductivity when the H_2-Pc+binder composite is "doped" by 5% C_{60}.

A series of UPS spectra is displayed in the left hand panel of Fig. 3 showing the valence states of different M-Pcs and the results for thin overlayers of C_{60} on these M-Pcs. For some of the M-Pcs a shift of the Pc-derived features towards the Fermi energy can be observed when they come in contact to C_{60}. This shift is attributed to a change in the surface-band bending [5]. Because of this shift, decomposition of the overlayer spectra into the single components is necessary in order to obtain the correct energy separation between the top of the valence band (VB) and the energy of the highest occupied molecular orbital (HOMO) of C_{60}. As a result the right panel shows the spectra of the individual M-Pcs with an energy scale that uses the top of the HOMO of C_{60} as a reference. The energy separation for the different M-Pc-C_{60} compounds ranges between 1.1 eV and 1.5 eV. The difference between this value and the gap energy of C_{60} (2.3 eV [7]) corresponds to the minimum energy necessary in order to transfer an electron from the occupied states of the M-Pc to the lowest unoccupied molecular orbital (LUMO) of C_{60}. This energy has to be supplied by the intermediate excitonic state. From optical transmission data [8] we know that H_2-Pc absorbs photons with energies higher than 1.5 eV. This corresponds well to the published data for other different M-Pcs where typical values range between 1.5 eV and 1.8 eV [1–9]. Since only 1.2 eV (in the case of α, β-H_2-Pc) are needed for the electron transfer, we can conclude that it is energetically possible. In the case of the other M-Pc-C_{60} compounds the situation is even better, because the required energy is smaller, as in the case of Cu-Pc where 0.8 eV are sufficient in order to transfer an electron from the VB of Cu-Pc into the LUMO of C_{60} as

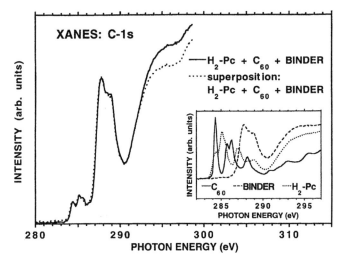

Fig. 2. Large frame: Measured XANES spectrum of a dispersion sample containing H_2-Pc, C_{60} and a polymeric binder (solid line) compared to a mathematical superposition (summation with weighting factors, broken line) of the measured spectra of the single components. inset: Measured XANES spectra of pure C_{60} (solid), pure binder (dashed) and pure H_2-Pc (dotted).

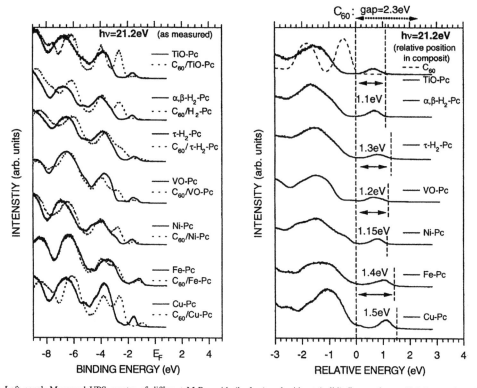

Fig. 3. Left panel: Measured UPS spectra of different M-Pcs with (broken) and without (solid) C_{60} overlayers (0.5–2 monolayers) on a binding energy scale. Right panel: Spectra of the single components (M-Pc: solid, C_{60}: broken) when the materials are in contact with each other on an energy scale referenced to the top of the HOMO of C_{60} as deduced from the spectra of the left panel.

can be directly concluded from Fig. 3. We therefore expect that all M-Pcs measured here will show an improved charge-generation efficiency upon doping with C_{60}.

The model in which the electron transfer reduces the recombination rate is proven by the fluorescence quenching in H_2-Pc and TiO-Pc displayed in Fig. 4. The fluorescence clearly decreases with increasing amounts of C_{60}. Since fluorescence and charge generation are competing processes, a quenched fluorescence indicates the improved charge-generation efficiency upon doping with C_{60}. In the case of H_2-Pc with 5% C_{60}, an improved photoconductivity has been measured compared to an undoped sample thereby demonstrating that the charge-generation efficiency is enhanced when the material is doped with C_{60} [8].

4. Conclusions

Using the methods of X-ray absorption near edge spectroscopy, ultra-violet photoelectron spectroscopy and fluorescence measurements we have demonstrated that C_{60} improves the light induced charge-generation efficiency of different metal-phthalocyanines. This effect is attributed to an electron transfer to the C_{60} that reduces the recombination rate of the primarily excited excitonic state. We thereby expect that these improved materials are promising candidates for application as photoreceptor materials in xerography of for laser printers.

Acknowledgements

We acknowledge financial support provided by the BMBF-VDI under contract number 13N6906. B.K. acknowledges financial support by the Ministerium für Wissenschaft und Forschung des Landes Nordrhein-Westfalen.

References

[1] Law K-Y. Chem Rev 1993;93:449.
[2] Sariciftci N, Heeger AJ. Int J Mod Phys 1994;B8:237.
[3] Watanabe A, Ito O. J Chem Soc, Chem Commun 1994;11:1285.
[4] Yoshino K, Akashi T, Yoshimoto K, Morita S, Sugimoto R, Zakhidov AA. Solid State Commun 1994;90:41.
[5] Schlebusch C, Kessler B, Cramm S, Eberhardt W. Synthetic Metals 1996;77:151.
[6] Material provided by: M Biermann, M Lutz, AEG Elektrofotografie, D-59581 Warstein-Belecke, Germany.
[7] Lof RW, van Venendaal MA, Koopmans B, Jonkman HT, Sawatzky GA. Phys Rev Lett 1992;68:3924.
[8] Schlebusch C, Morenzin J, Kessler B, et al. In: Kafafi ZH, editor. Fullerenes and photonics IV. Proceedings of SPIE 1997; 3142, 120.
[9] Kakuta A, Mori Y, Takano S, Sawada M, Shibuya I. J Imag Technol 1985;11:7.

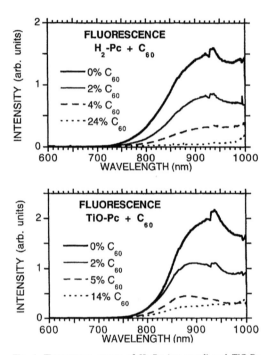

Fig. 4. Fluorescence spectra of H_2-Pc (top panel) and TiO-Pc (bottom panel) samples containing different amounts of C_{60}.

Pergamon

Carbon 37 (1999) 721–725

CARBON

Gd–L$_{III}$ EXAFS study of structural and dynamic properties of Gd@C$_{82}$ between 10 and 300 K

H. Giefers[a], F. Nessel[a], S.I. Györy[a], M. Strecker[a], G. Wortmann[a,*],
Yu.S. Grushko[b], E.G. Alekseev[b], V.S. Kozlov[b]

[a]*Universität-GH Paderborn, Fachbereich Physik, D-33095 Paderborn, Germany*
[b]*St. Petersburg Nuclear Physics Institute, Gatchina 188350, Russia*

Received 16 June 1998; accepted 3 October 1998

Abstract

The location of endohedral Gd ions in Gd@C$_{82}$ is studied by Gd–L$_{III}$ X-ray absorption spectroscopy in the temperature range 10–300 K. The near-edge data point to a covalent bond of trivalent Gd ions with the carbon cage. The EXAFS data can be well fitted with a Gd position above a carbon hexagon with two neighbour shells of carbon at distances $R_{1,2}=2.49(3)$ and 2.95(5) Å, corresponding to a large off-center position of about 1.8 Å from the center of the C$_{82}$ cage. The second cumulants of these distances are relatively large and weakly depending on temperature. This points to a considerable distribution in Gd–C$_1$ and Gd–C$_2$ distances and confirms the relatively strong binding of the Gd ion to the C$_{82}$ cage. These results will be discussed together with previous EXAFS studies and theoretical calculations of Y@C$_{82}$ and La@C$_{82}$. © 1999 Elsevier Science Ltd. All rights reserved.

Keywords: Endohedral fullerenes; C. EXAFS

1. Introduction

The location of endohedral R ions (R=lanthanides, Y) inside the fullerene cages and their dynamic behaviour is of actual interest [1–3]. In the standard production process by co-evaporation of graphite and R oxides, the most abundant soluble endohedral species are R@C$_{82}$ molecules. A large number of theoretical and experimental studies were performed to clarify which modifications (isomers) of C$_{82}$ and R@C$_{82}$ are energetically preferable and therefore most abundant [1,4–7]. There is strong experimental and theoretical evidence for a considerably off-center location of the R ions from the center of the C$_{82}$ molecule, since the inner diameter of the C$_{82}$ cage is much larger than the size of trivalent R ions; this is also due for divalent R ions as in the case of Tm@C$_{82}$ [3]. The most direct experimental determination of the location in the cage is provided by X-ray absorption studies at the R ions. The extended X-ray absorption fine structure (EXAFS) can give detailed information about the number and distances

of the nearest and next-nearest carbon atoms [8,9]. Such a study was first performed at the Y K-edge of Y@C$_{82}$, proving the endohedral nature of the Y ions, located inside the cage above a hexagon with 6 carbons at a nearest distance of 2.40(5) Å and a second nearest carbon shell at 2.85(5) Å. A more recent La–L$_{III}$ EXAFS study of La@C$_{82}$ came to similar conclusions about the endohedral location of La [9].

Here we present a Gd–L$_{III}$ X-ray absorption study of Gd@C$_{82}$, performed in the temperature range 10–300 K. From a detailed analysis of the EXAFS structures, we found a location for the Gd ions similar to that of Y and La in previous EXAFS studies. From the temperature variation of the second cumulant of the Gd–C distances, $\sigma^2(T)$, also called the EXAFS Debye–Waller factor, we can distinguish between dynamic and static contributions. The former results from thermal vibrations, the latter from local distortions or distributions of distances within a neighbour shell. We found a relatively large static distribution of Gd–C distances in the first and second neighbour shell, reflecting the variance of distances in the distorted hexagons of the possible C$_2$ and C$_{2v}$ isomers of

*Corresponding author

0008-6223/99/$ – see front matter © 1999 Elsevier Science Ltd. All rights reserved.
PII: S0008-6223(98)00261-9

the Gd@C$_{82}$ molecule. From the observed weak temperature dependence of $\sigma^2(T)$ we derive a relatively strong
binding of the Gd(3+) to the C$_{82}$ cage. The latter finding
is supported by the near-edge structure (XANES) of the
Gd–L$_{III}$ edge.

2. Sample preparation and experimental details

2.1. Sample preparation

Graphite anodes impregnated with Gd$_2$O$_3$ in a C/Gd
(at.) ratio of ~100 were evaporated in an 70 A DC arc at a
He pressure of 300 mbar. At that He pressure the highest
production yield for Gd endohedrals is obtained, amount-
ing to 15 wt.% from the extractable fullerenes [10]. Details
of the solvent extraction of the endohedral fullerenes from
the soot will be described elsewhere. The elemental
analysis of Gd and C of the carbonaceous materials was
performed by Rutherford backscattering. Thermal and laser
desorption TOF mass spectra of the extracted endohedral
fullerenes reveal the presence of the Gd@C$_{82}$ isotopic
multiplets. The composition of the present sample, according to elemental and HPLC analysis, was (in wt.%) 36%
Gd@C$_{82}$, 24% C$_{60}$, 27% C$_{70}$, 7% C$_{76,78}$ and 6% C$_{82,84}$.
We could detect with the present resolution no Gd@C$_{80}$,
which is known to be produced by this preparation route in
small amounts (less than 2% of Gd@C$_{82}$ [11]). In the
following the sample is named Gd@C$_{82}$, since only the
properties of Gd will be investigated. The amount of
Gd@C$_{82}$ used in the EXAFS study was about 30 mg.

We should state at this point that there is ample evidence
that the local structural and electronic properties of the
endohedral Gd ions are not (or only very little) influenced

by the fact that endohedral fullerenes are surrounded in the
sample by empty fullerenes with weak van der Waals
interactions between them. The above cited EXAFS study
of Y@C$_{82}$ [8] was performed as in the present case with a
mixture of endohedral and empty fullerenes; the derived
Y–C distance in Y@C$_{82}$ agrees very well with that of a
recent XRD diffraction study of a pure Y@C$_{82}$ crystal [2]
and with theoretical calculations [4–7]. An XPS study
came to the conclusion that the electronic properties of
R@C$_{82}$ are the same as a pure solid or in mixture with
empty endohedrals [6]. All theoretical calculations consider only isolated R@C$_{82}$ molecules.

2.2. XAS experiments

The X-ray absorption studies were performed at the
EXAFS-II beamline of HASYLAB (DESY, Hamburg)
employing a Si(1,1,1) double crystal monochromator and a
focusing mirror (providing also a suppression of the higher
harmonics). The sample was encapsulated in a plastic
holder with 5 mm in diameter and placed in the exchange
gas of a He cryostat. The monochromatized synchrotron
radiation was measured before and behind the sample with
two ionization chambers filled with N$_2$. A GdF$_3$ absorber
was measured simultaneously as reference absorber behind
the Gd@C$_{82}$ sample with a third ionization chamber.

3. Results and discussion

The Gd–L$_{III}$ spectra were measured from 7100 eV to
7950 eV. A typical experimental spectrum is shown in Fig.
1. It exhibits dominant resonances ("white-lines") at the
L$_{III}$ and L$_{II}$ thresholds, which amplitudes are about three

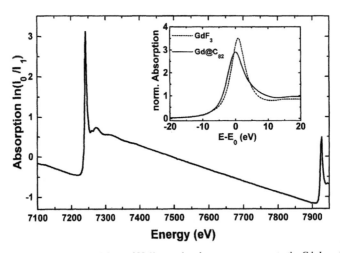

Fig. 1. Gd–L$_{III}$ X-ray absorption spectrum of Gd@C$_{82}$ at 300 K, covering the energy range up to the Gd–L$_{II}$ edge. The inset shows the
normalized Gd–L$_{III}$ near-edge region of Gd@C$_{82}$ and a GdF$_3$ reference absorber.

times larger than the normalized absorption height. From the energy position and comparison with the GdF_3 reference absorber (see inset) it follows immediately that the Gd ions are present in a trivalent state. Interestingly the white-line of $Gd@C_{82}$ is lower in amplitude and broader than that of GdF_3 and shifted by about 0.8 eV to lower energies. This points to stronger crystal field interactions and, accordingly, to a stronger covalent bond for the $Gd@C_{82}$ case. Details of the analysis of the $L_{I\text{-}III}$ edges of $Gd@C_{82}$ will be given in a forthcoming paper.

The EXAFS oscillations, starting about 30 eV above the edge, are very weak, even at the lowest temperature. This is due, on the one hand, to the weak scattering strength of carbon as a light element, which decreases also quite rapidly for higher k values of the excited electrons. The other reason is, as discussed below, the relatively low coordination number of carbon neighbours and distance distributions in carbon shells. It is worth mentioning in this context that the most favourable case of a well-defined and highly coordinated carbon surrounding (with NC=24) is given by alkali ions located on the tetrahedral sites of "exohedral" A_3C_{60} and A_2BC_{60} (A=K, Rb; B=Rb, Cs) systems [12].

Using standard evaluation procedures [12,13], the Gd–L_{III} EXAFS oscillations, $\chi(k)$, were extracted from the raw data up to $k=9$ Å^{-1} (at higher k values, the present data are too noisy to be used for an EXAFS evaluation). Great care was taken to perform identical background removals for all spectra of the temperature series. As shown in Fig. 2(a), there is only a weak temperature dependence between the $\chi(k)$ spectra at 10 K and 300 K. It should be mentioned at this point that the $\chi(k)$ spectra of $La@C_{82}$ and $Y@C_{82}$ are very similar in their principal features to that of $Gd@C_{82}$ (the Y K-edge spectrum has a phase shift of 180° in comparison the L_{III} spectra of La and Gd). The strong multi-electron excitation visible in the La $\chi(k)$ spectra around 5.7 Å^{-1} [9] has a much weaker correspondence in the present Gd $\chi(k)$ spectra at 6.3 Å^{-1}. This disturbance of the EXAFS signal has been deglitched before the following analyses.

Figure 2(b) shows the Fourier transforms of the above $\chi(k)$ spectra obtained in the range from 2.6 to 9 Å^{-1}. There is only one (somewhat broadened) peak A at a (phase-shifted) distance of $R'=2.0$ Å and no resolved structures at higher R' values. The relatively small change with temperature in the amplitude of peak A reflects again the weak temperature dependence of the EXAFS oscillations. A range of $R'=1.3$–2.5 Å was used for the backtransformed $\chi(k)$ spectra, shown in Fig. 2(c) with a two-shell fit described below.

In our fits of the backtransformed $\chi(k)$ spectra we followed the lines described in [8,9]. For the calculations of scattering amplitudes and phase shifts we used the FEFF6.01 program [14,15], which was already successfully applied in a previous version (FEFF5.05) for the EXAFS evaluation of exohedral fullerene systems [12]. A

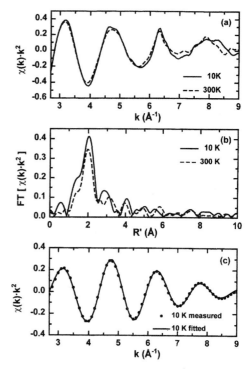

Fig. 2. (a) Gd–L_{III} EXAFS oscillations, $\chi(k) \cdot k^2$, of $Gd@C_{82}$, at two temperatures; (b) magnitude of Fourier transforms of the above k^2 weighted $\chi(k)$ spectra; (c) backtransformed k^2 weighted $x(k)$ spectrum with a two-shell analysis.

fit with only a single shell of Gd–C distances to the backtransformed $\chi(k)$ spectra was not successful. Fits with two Gd–C shells were more successful; these fits require seven parameters: E_0, the threshold energy, N_1 and N_2, the number of neighbours, R_1 and R_2, the Gd–C distances and finally σ_1^2 and σ_2^2 the second cumulant of the distances. We used several parameter sets with N_1 and N_2 values fixed to various models for carbon surroundings in the C_{82} cage [8,9]. The best set of parameters was found with N_1 and N_2 around six. A second round of fits of all spectra of the temperature series was performed with $N_1=N_2=6$ and a fixed E_0 value with $R_{1,2}$ and $(\sigma^2)_{1,2}$ as remaining free (and less correlated) parameters. The fitted results for the Gd–C distances $R_{1,2}$ and the $\sigma_{1,2}$ values are shown in Fig. 3. The first Gd–C shell has a distance of 2.49(3) Å and the second shell a distance of 2.95(5) Å at 10 K, both independent of temperature within the relative accuracy of 0.01 Å. The $(\sigma^2)_{1,2}$ values are already relatively large at 10 K and exhibit a modest increase with temperature.

Before we discuss our results, we want to mention that our $R_{1,2}$ and $(\sigma^2)_{1,2}$ values agree quite well with those obtained in the above cited $Y@C_{82}$ EXAFS study at 10 K [8] and, to a less extent, with those of the $La@C_{82}$ study

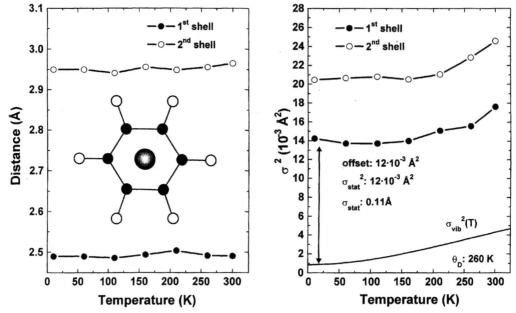

Fig. 3. Temperature dependence of the derived parameters of a two-shell fit to the backtransformed $\chi(k)$ spectra of Gd@C$_{82}$: (left) Gd–C distances R$_1$ and R$_2$ of the first and second carbon shell, shown schematically in the inset; (right) temperature dependence of the second cumulants $(\sigma^2)_{1,2}$ with an analysis in terms of static and dynamic contributions.

obtained at 300 K [9]. In this latter study a distance of the first La–C shell of 2.47(3) Å was derived, slightly smaller than the present Gd–C$_1$ value; from the ionic size of La(3+) one expects a larger value, as actually found in theoretical studies as 2.54 Å [6] and 2.65 Å [7]. Our Gd–C$_1$ value agrees also favourably with Y–C$_1$ derived from the XRD study of Y@C$_{82}$ [2].

The $\sigma_{1,2}$ values of 0.10 and 0.13 Å given in the Y EXAFS study [8] as well as of 0.11 and 0.15 Å given in the La EXAFS study [9] must be squared to be compared with our $(\sigma^2)_{1,2}$ values in Fig. 3.

We analyse now the temperature dependence of the second cumulant considering both static and thermal disorder: $\sigma^2(T) = \sigma_{stat}^+ \sigma_{vib}^2(T)$. With a simple Debye model, we derive from the temperature variation a local Debye temperature of 260(20) K. There remains a large static contribution of about $12 \cdot 10^{-3}$ Å^2, corresponding to a variation in distances by 0.11 Å.

This variation in distances can be attributed to three reasons: (i) About half of it originates from the variance in Gd–C distances due to the fact that the carbon hexagons are distorted (more in the C$_2$ than in the C$_{2v}$ and C$_{3v}$ isomers). Even more important, the calculation in [6] show that these distortions are increased by the insertion of R(3+) ions, causing an increase of the neighbouring C–C bonds and a larger variation of the C–C and R–C distances. (ii) Different Gd@C$_{82}$ isomers, if present in the

sample, may have slightly different local cage structures, thus enlarging the value of σ_{stat}^2. (iii) The Gd ions might be slightly displaced from a centered position over the hexagons; here a small displacement by 0.05 Å could easily account for the observed large value of σ_{stat}^2.

Finally we want to comment on the fact that we could not observe in a Gd-155 (86.5 keV) Mössbauer study of the present Gd@C$_{82}$ sample a resonance signal larger than 0.02% at 4.2 K. From the known absorber thickness and the source properties, we can derive, even in the case of paramagnetic broadening of the resonance, an upper limit of the recoil-free fraction (f-factor) as 0.01 at 4.2 K. Applying the Debye model, this value converts to a Debye temperature of Θ_D(ME)=100 K, which is drastically different from the above value of Θ_D(EXAFS). In contrast to EXAFS, which monitors the near intramolecular surrounding of the Gd ion, the Mössbauer effect is sensitive to the properties of the whole crystal, here a molecular solid with weak van der Waals bonds between the (endohedral and empty) fullerenes. The large recoil energy of 26 meV can excite intermolecular vibrations and, due to the off-center location of the Gd ion, also very easily molecular rotations of the Gd@C$_{82}$ entity. This case will be discussed in detail in an other contribution of this symposium, reporting on the observation of the Dy-161 (25.7 keV) Mössbauer resonance in Dy@C$_{82}$ [16]; here the lower gamma energy with a much smaller recoil energy of 2.2

meV is more favourable for a Mössbauer study of endohedral systems.

4. Conclusion

The present EXAFS study derives a position of the trivalent Gd ion above the carbon hexagon; the Gd–C distances agree well, considering the different size of the R(3+) ions, with previous experimental and theoretical studies. We observe, as in the other EXAFS studies, a relatively large variance in these distances. We derive from the weak temperature dependence of the second cumulants a relatively strong binding of the Gd(3+) to the C_{82} cage. Such a strong covalent bond is also reflected by the Gd L_{III}-edge structure. Future EXAFS studies will be devoted to the Dy@C_{82} system, where preliminary Dy–L_{III} studies show a very similar spectral behaviour. Combination of EXAFS and Mössbauer results should provide a detailed picture of the structural and dynamic properties of R@C_{82} systems.

Acknowledgements

This work was supported by the Deutsche Forschungsgemeinschaft (grants Wo209/10 and 436 RUS 17/162/95) and by the Russian Foundation for Intellectual Collaboration.

References

[1] Kuzmany H, Fink J, Mehring M, Roth S, editors. Fullerenes and Fullerene Nanostructures. Singapore, World Scientific, 1996.

[2] Takata M, Umeda B, Nishibori E, Sakata M, Saito Y, et al. Nature 1995;377:46, (Ref. 1, p. 155 (the latter reference should be used for Y–C distances)).

[3] Pichler T, Golden MS, Knupfer M, Fink J, Kirbach U, et al. Phys Rev Lett 1997;79:3026.

[4] Laasonen KE, et al. Science 1992;258:1916, (Andreoni W, Curioni A, in Ref. 1, p. 205).

[5] Nagase S, Kobayashi K, Akasaka T. In: Kuzmany H, Fink J, Mehring M, Roth S, editors. Fullerenes and Fullerene Nanostructures. Singapore, World Scientific, 1996:161.

[6] Poirier DM, Knupfer M, Weaver JH, Andreorie W, Laasonen K, et al. Phys Rev B 1994;49:17403.

[7] Kobayashi K, Nagase S. Chem Phys Lett 1998;282:235.

[8] Park C-H, Wells BO, DiCarlo J, Shen Z-X, Salem JR, et al. Chem Phys Lett 1993;213:196.

[9] Nomura M, Nakao Y, Kikuchi K, Achiba Y. Physica B 1995;208–209:539.

[10] Alekseev EG, Karataev VI, Kozlov VS, Lebedev VM, Khodorkovskii MA, et al. Proceedings of the International Workshop on Fullerenes and Atomic Clusters. St. Petersburg, 1997, in print. (See also Ref. 15, where a similar preparation route was used for Dy@C82).

[11] Funasakai H, Sakurai K, Oda Y, Yamamoto K, Takahashi T. Chem Phys Lett 1995;232:273.

[12] Nowitzke G, Wortmann G, Werner H, Schlögl R. Phys Rev B 1996;54:13230.

[13] Teo BK. EXAFS: Basic Principles and Data Analysis. Berlin: Springer–Verlag, 1986.

[14] Rehr JJ, Mustre de Leon J, Zabinsky SI, Albers RC. J Am Chem Soc 1991;113:5135.

[15] Mustre de Leon J, Rehr JJ, Zabinsky SI, Albers RC. Phys Rev B 1991;44:4146.

[16] Grushko YuS. Unpublished results.

Pergamon

Carbon 37 (1999) 727–732

CARBON

C$_{60}$ thin films on Ag(001): an STM study

G. Costantini*, S. Rusponi, E. Giudice, C. Boragno, U. Valbusa

INFM-Unità di Ricerca di Genova, CFSBT-CNR, Dipartimento di Fisica dell'Università di Genova, via Dodecaneso 33, 16146 Genoa, Italy

Received 16 June 1998; accepted 3 October 1998

Abstract

The structure of submonolayer C$_{60}$ films deposited on Ag(001) is studied as a function of the substrate temperature in the range 100–700 K. Morphological aspects such as island shape and preferential nucleation sites are in agreement with the theory of submonolayer deposition. A peculiar irreversible transition which leads to the onset of brightness differences between molecules of the same film is observed at around 300 K. This effect is ascribed to electronic differences due to non-equivalent orientation of the C$_{60}$ molecules on the surface rather than to morphological differences induced by a surface reconstruction. © 1999 Elsevier Science Ltd. All rights reserved.

Keywords: A. Fullerene, Carbon films; C. Scanning tunneling microscopy (STM)

1. Introduction

Great interest has recently focused on the technological and scientific applications of fullerene-based materials. Promising characteristics of these systems ranging from mechanical or tribological aspects such as lubrication and coating [1] to optical and electronic properties [2] have been investigated. The discovery of high-T_c superconductivity in alkali-doped fullerenes [3] has directed great efforts towards the study of C$_{60}$–metal interactions and particularly of C$_{60}$ thin films grown on metal substrates. The conviction that electronic together with many other interesting physical properties strongly depend upon the crystalline structure of these films and on their interaction with the surface, gives importance to the understanding and control of the growth mechanisms. In this field, scanning tunnelling microscopy has been revealed as an important tool for the study of films on both metal and semiconductor substrates since it allows a detailed analysis of the nucleation and growth processes.

The deposition of submonolayer fullerene films on silver has already been studied from an electronic point of view. In particular Chase et al. [4] found that among the

polycrystalline noble metal substrates silver presents the strongest interaction with the C$_{60}$ overlayer which is characterised by its metallic nature and by a large charge transfer from the substrate. For single crystal Ag(001) substrates the importance of this interaction seems to increase, as recently reported by Goldoni et al. [5]: a particularly high charge transfer is accompanied by a considerable reduction of the density of states at the Fermi level, corresponding to the opening of a pseudogap, at about 120 K.

In this paper we present STM results for the growth of submonolayer C$_{60}$ films on Ag(001) over a wide temperature range (100–700 K). The most relevant differences we observed, passing from lower to higher temperatures, concern the structures of C$_{60}$ islands together with the preferential nucleation sites and the onset of a peculiar bright–dim contrast between molecules of the same film. The change with temperature of the former characteristics is in qualitative agreement with the theory of submonolayer deposition (for a review see, for example, chapter 17 in [6] and references therein). On the other hand the appearance of bright and dim C$_{60}$ molecules, which occurs also for other substrates [7], and the particular structures in which these molecules organise, have not been yet completely explained. The bright–dim contrast appears as the result of an irreversible temperature transition that takes place around 300 K. We attribute this

*Corresponding author. Tel.: +39-10-353-6356; fax: +39-10-311-066.
E-mail address: costantini@ge.infn.it (G. Costantini)

0008-6223/99/$ – see front matter © 1999 Elsevier Science Ltd. All rights reserved.
PII: S0008-6223(98)00262-0

contrast to electronic differences due to non-equivalent orientations of the adsorbed molecules, rather than to topographic differences induced by a surface reconstruction.

2. Experimental

The experimental apparatus consists of a variable temperature UHV-STM equipped with standard facilities. It is possible to have full temperature control during each stage of the sample preparation, C_{60} deposition and subsequent imaging and, thanks to a particularly compact design, no sample transfer is needed, so that measurement can start immediately after deposition. More details can be found in Ref. [8]. The Ag(001) crystal was prepared with several cycles of 1 keV Ar$^+$ sputtering followed by annealing up to 750 K; this procedure routinely produced terraces with an extension of at least 600 Å. C_{60} was deposited heating at 620 K a Ta crucible filled with 99.9% purity powder. During deposition the substrate was kept at a fixed temperature T_d and in some cases after deposition it was post-annealed at temperature T_a. The deposition rate was fixed at around 1 ML/min while coverage was changed varying the exposure time. The reported coverages were estimated by direct analysis of the statistics of STM topographies. All images have been recorded in constant current mode with typical tunnelling parameters $I=1$ nA and $V=\pm1.5$ V and, in order to enhance the contrast, are often presented in derivative mode, thus appearing as illuminated from the left hand side.

3. Results and discussion

Comparing the Ag(001) atomic structure with complete C_{60} monolayers grown at high temperatures (T_d or $T_a>450$ K) we found that the fullerene molecules organise themselves in a c(6×4) superlattice for which every site is equivalent, with a nearest neighbour (NN) distance of 10.42 Å (10.02 Å in bulk C_{60} [9]) derived taking an Ag lattice constant of 4.09 Å (Fig. 1). The square symmetry of the silver substrate allows for two equivalent domains mutually rotated by 90° that have been actually observed. These results are in agreement with recent LEED measures on the same system in the higher temperature regime [5] ($T_d=750$ K).

At low coverages, in the low temperature regime ($T_d=100$ K), C_{60} molecules appear to nucleate preferentially at steps or dislocations (both on the upper as on the lower terrace) but also on flat terraces (Fig. 2a). We observed the formation of little islands characterised by a quasi-fractal shape: their area–perimeter ratio is low and they are delimited by a great number of short straight edges (less than three molecules per edge) oriented along the closed packed directions. The fact that these islands do not show a

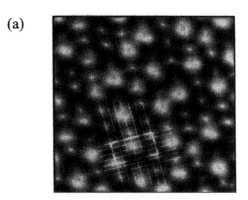

(a)

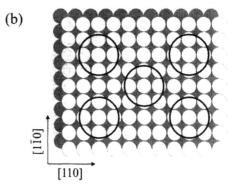

(b)

Fig. 1. (a) Large C_{60} island at high resolution; the marked grid represents the underlying silver lattice. (b) Model representation of the C_{60} superlattice.

complete fractal nature similar to that obtained in the DLA growth model [6], indicates that at 100 K there is still some diffusion of the molecules along the C_{60} island edges. It is noteworthy that, at least in the submonolayer regime, the island structure is consistent, within the experimental errors, with the described c(6×4) C_{60} superlattice structure over the whole temperature range covered, which means also in films grown at temperatures down to 100 K.

Figures 2b–e have been taken on the same film as Fig. 2a, but with a post-annealing temperature of, respectively, $T_a=150$, 200, 250 and 300 K. It is evident that upon increasing the substrate temperature, the islands increase in extent and assume a more compact structure, while their edges become straighter and longer, until for $T_a>250$ K their shape is almost hexagonal. These modifications are due to a temperature-induced increase of the diffusion along the island edges which smooths out the perimeter over short length scales.

Increasing the deposition temperature T_d, we observed an increasing tendency towards nucleation at step edges.

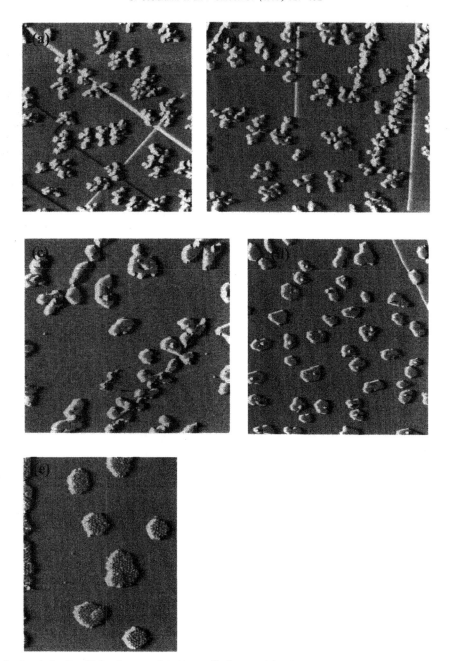

Fig. 2. (a) C_{60} deposited at $T_d = 100$ K and not annealed; (b) same film but annealed at $T_a = 150$ K; (c) $T_a = 200$ K; (d) $T_a = 250$ K; (e) $T_a = 300$ K.

The island density on terraces decreases especially on narrow terraces while their shape becomes more compact until, for $T_d > 300$ K nucleation takes place exclusively at step edges, since for these temperatures the mobility of C_{60} molecules on Ag(001) is high enough for their mean free path to be larger then the mean terrace width.

The preferential nucleation at step edges for C_{60} films deposited or annealed at high temperatures has been reported also for other noble metal substrates such as Ag(111) [10], Au(111) [11] and Cu(111) [12], although in the case of Ag(110) [13] and Cu(110) [14] the probably stronger interaction of C_{60} molecules with these more open surfaces leads to a homogeneous nucleation on the surface, making terrace adsorption sites favourable also at higher temperatures (T_d=300 K, T_a=650 K for [13] and T_d=600 K for [14]).

At higher coverages and in the higher temperature regime (T_d>300 K) the growth proceeds outwards from the completely saturated steps and large C_{60} islands gradually extend over the terraces. At room temperature a further increase in the coverage gives rise to the addition of C_{60} molecules to the second layer before the first is completed (Fig. 3a). The exact temperature at which this phenomenon sets in and the percentage of molecules that belong to the second layer strongly depend on the deposition rate. For temperatures higher than about 500 K, independently of the coverage and deposition rate, no second layer was observed, meaning that all molecules, except those directly bound to the silver substrate, desorb. This result is in good agreement with measurements done by Goldoni et al. [5] who estimated a second layer desorption temperature in the 500 K range also.

Another interesting feature emerges from the inspection of Figs. 2a–e, namely that for temperatures below 300 K all C_{60} molecules are equivalent, while at about 300 K some molecules begin to appear brighter than others (Fig. 2e). This bright–dim contrast is present in all films deposited or annealed above 300 K and becomes much more evident with increasing annealing temperatures [15]. Moreover this transition appears to be irreversible, i.e. cooling down a film that has been grown or post-annealed at temperatures higher than 300 K does not eliminate the contrast between molecules. The brightness of each molecule remains constant over periods of hours and does not depend on the sign of the tunneling potential since images taken over the same region at opposite gap polarisations are identical.

In STM images, brighter features usually correspond to higher topographical features. In our case the (apparent) height difference between bright and dim molecules turns out to be ~2 Å, much too low to represent a C_{60} molecule of the second layer since the calculated C_{60} cage diameter is 7.1 Å [16] and the step height in the (111) surface of bulk fcc C_{60} in known to be 8.1 Å [9]. Moreover the brighter molecules appear always in lattice positions, which would not be expected from second layer molecules.

The bright molecules are not arranged in a random manner but organise themselves in ordered structures. Figures 3a and 3b have been taken on C_{60} films both deposited at room temperature, but the film in Fig. 3b has been post-annealed at 670 K. It appears clearly that in the non-annealed film there are many different structures

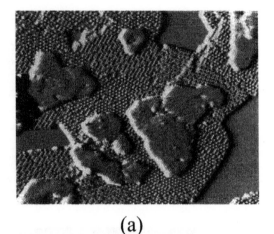

(a)

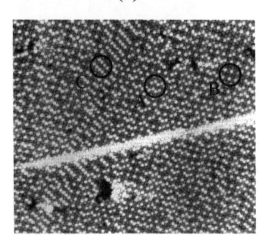

(b)

Fig. 3. (a) 1.1 ML of C_{60} deposited at T_d=300 K (scan area ~1200×800 Å^2). (b) 1.1 ML of C_{60} deposited at T_d=300 K and annealed at T_a=670 K (scan area ~520×600 Å^2); refer to the text for evidenced structures A, B and C.

extending over small domains of around 200×200 Å^2, whereas for T_a=670 K the C_{60} overlayer appears much more ordered. For high annealing temperatures (T_a>450 K) the bright molecules represent ~18% of the total number and only three fundamental structures (evidenced in the marked circles A, B and C of Fig. 3b) are visible and extend over the whole film. Structure A, which is also the most frequent one, is a zigzag along the [−130] direction of the silver substrate; curiously it never appears in the equivalent [−310] direction. B is a linear structure along the same direction (never along $[-310]$) which can thus be obtained from A getting dim every second molecule in the zigzag, while structure C corresponds to the line boundary of two A structures with a mutual shift perpendicularly to the zigzag direction.

These structures can be obtained as the superposition of suitably shifted copies of a simple lattice of C_{60} molecules with a square unit cell composed by one dim and three bright molecules. Fig. 4 shows how to recover the observed structures A, B and C.

In STM measurements it is not possible to distinguish whether or not a brighter feature that corresponds to a local increase in the current value is due to the reduction of the tunneling gap, and thus to an increased local height of the surface, or to an increase of the local density of states and thus to an electronic effect. Nevertheless in our case many arguments brought us to the conviction that the brightness contrast observed between molecules of the same film is due to an electronic rather than to topographic effects. The main arguments against surface reconstruction originate from experimental observations as the intermolecular distance, which is close to the C_{60} bulk value and thus should minimise mechanical stress, the general flatness of the second layer islands and the great number of different structures extending over small domains observed at room temperature grown films (Fig. 3a) that can be explained only by a very complicated reconstruction. In order to account for the described phenomena we proposed a model in which the molecules, deposited at temperatures below the measured transition temperature (~ 300 K), are not chemically bound to the surface, possibly rotating, and

thus, thanks to an average effect, give a uniform aspect to the films they compose. For higher temperatures the charge transfer sets in from the substrate to the C_{60} molecules belonging to the first layer which thus bind in different orientations in respect to the surface. This situation generates local differences in the density of states (for example due to pentagonal or hexagonal C-rings facing the vacuum) and thus can explain the brightness differences. A more detailed discussion can be found elsewhere [15].

Preliminary results from XPD measures on the same system seem to confirm some aspects of the presented model [17]. Similar conclusions about the bright–dim contrast were reached also for the C_{60}–Ag(111) and C_{60}–Au(111) systems by Altman and Colton [7].

4. Conclusions

In conclusion we report the temperature dependence of C_{60} submonolayer films in the range 100–700 K. We observed nucleation and growth characteristics that follow the predictions of usual growth theory. In addition we observed a temperature transition which takes place around 300 K evidenced by the onset of brightness differences between molecules of the same film. We propose a qualitative model which attributes these differences to electronic effects and in particular to non-equivalent adsorption orientations of the C_{60} molecules due to the temperature-dependent onset of charge transfer from the surface.

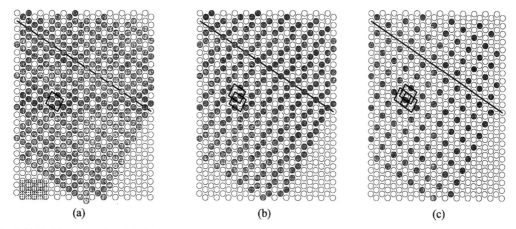

(a)   (b)   (c)

Fig. 4. (a) Model representation of the fundamental square lattice corresponding to the linear pattern B in Fig. 3b. The upper and the lower part are mutually displaced by one NN distance along the marked closed packed direction. (b) Superposition of one shifted copy of the fundamental lattice: the zigzag pattern A appears together with the linear pattern C in the separation region between the two displaced lattices. (c) Superposition of two shifted copies of the fundamental lattice: the resulting pattern, made up by one dim molecule surrounded by six bright molecules, corresponds to one of those observed in not annealed films (upper right part of Fig. 3a).

Acknowledgements

We acknowledge useful and fruitful discussion with M. Sancrotti and C. Cepek to whom in particular goes our gratitude for providing preprints and preliminary experimental results. E. Magnano contributed in many aspects to the present experiment. This work was supported by INFM under the project CLASS-PRA.

References

[1] Brenner DW, Harrison JA, White CT, Colton RJ. Thin Solid Films 1991;206:220.

[2] Wang Y. Nature 1992;356:585.

[3] Hebard AF, et al. Nature 1990;350:600.

[4] Chase SJ, Bacsa WS, Mitch MG, Pilione LJ, Lannin JS. Phys Rev B 1992;46:7873.

[5] Goldoni A, Cepek C, Magnano E, Laine AD, Vandre' S, Sancrotti M. Unpublished, 1997.

[6] Barabasi AL, Stanley HE. Fractal concepts in surface growth. Cambridge: Cambridge University Press, 1995.

[7] Altman EI, Colton RJ. Phys Rev B 1993;48:18244.

[8] Conti R, Rusponi S, Pagnotta D, Boragno C, Valbusa U. Vacuum 1997;48:639.

[9] Heiney PA, Fisher JE, McGhie AR, Romanow WJ, Denenstein AM, McCauley JP, Smith AB, Cox DE. Phys Rev Lett 1991;66:2911.

[10] Altman EI, Colton RJ. Surf Sci 1993;295:13.

[11] Altman EI, Colton RJ. Surf Sci 1992;279:49.

[12] Hashizume T, Motai K, Wang XD, Shinohara H, Saito Y, Maruyama Y, Ohno K, Kawazoe Y, Nishina Y, Pickering HW, Kuk Y, Sakurai T. Phys Rev Lett 1993;71:2959.

[13] David T, Gimzewski JK, Purdie D, Reihl B, Schlittler RR. Phys Rev B 1994;50:5810.

[14] Pedersen MØ, Murray PW, Lægsgaard E, Stensgaard I, Besenbacher F. Surf Sci 1997;389:300.

[15] Giudice E, Magnano E, Rusponi S, Boragno C, Valbusa U. Surf Sci 1998;405:561.

[16] Kroto H. Nature 1988;242:1139.

[17] Cepek C. Personal communication, 1998.

Pergamon

Carbon 37 (1999) 733–738

CARBON

Electron energy-loss spectroscopy studies of single wall carbon nanotubes

M. Knupfer[a,*], T. Pichler[a], M.S. Golden[a], J. Fink[a], A. Rinzler[b], R.E. Smalley[b]

[a]*Institut fur Festkörper- und Werkstofforschung Dresden, Postfach 270016, D-01171 Dresden, Germany*
[b]*Center for Nanoscale Science and Technology, Rice Quantum Institute, Departments of Chemistry and Physics, Ms-100, Rice University, P.O. Box 1892, Houston, TX 77251, USA*

Received 16 June 1998; accepted 3 October 1998

Abstract

We have carried out momentum-dependent measurements of the density response function of bulk samples of purified single wall nanotubes using electron energy-loss spectroscopy. Carbon nanotubes support both excitations between delocalized and localized electronic states. The π-plasmon exhibits significant q-dependence, with a dispersion relation similar to that of the graphite plane, demonstrating the graphitic nature of the nanotube electron system along the tube axis. In contrast, the interband excitations observed at low energy have vanishingly small dispersion in q. These excitations between localized states are related to characteristic interband transitions between the singularities in the nanotube electronic density of states, and can thus be used to show that our samples contain significant quantities of both semiconducting and metallic single wall nanotubes. © 1999 Elsevier Science Ltd. All rights reserved.

Keywords: A. Carbon nanotubes; C. EELS; D. Electronic properties; Optical properties

1. Introduction

Carbon nanotubes are a new member of the growing family of novel fullerene-based materials, and represent a promising extension of this material class as they can be used as ideal building blocks for nanoengineering as a result of their special electronic [1,2] and mechanical [3] properties. Regarding their molecular structure, nanotubes can be envisaged as rolled-up graphene sheets which are capped with fullerene-like structures. Interestingly, their electronic properties are predicted to vary depending upon the wrapping angle and diameter of the graphene sheet, thus offering the possibility to selectively form either metallic or semiconducting nanotubes [4–6].

The best system in which to investigate the intrinsic properties of this new material class are single wall nanotubes (SWNTs). Macroscopic nanotube samples generally contain a distribution of tubes with different diameters and chirality and thus present the experimentalist with an averaged picture of their properties. Therefore, many studies have been carried out on individual SWNTs. Transport measurements [7] and scanning tunnelling

spectroscopic (STS) and topographic (STM) studies of *single* nanotubes [8] have done much in the recent months to advance our knowledge regarding the properties of SWNTs, for example by experimentally verifying the remarkable relationship between nanotube geometry and their electronic properties [8]. In addition, spatially-resolved electron energy-loss spectroscopy (EELS) has been performed on individual multi-wall nanotubes (MWNT) [9–11] or on a single bundle of SWNTs [12].

Much less has been done using methods that can be applied to macroscopic samples. Combined electron spin resonance, microwave and DC resistivity measurements [13] have led to the conclusion that bulk SWNT material is metallic. Resonant Raman measurements have been interpreted in terms of the vibrational modes of non-chiral 'armchair' SWNTs [14], which is in contrast to the results from scanning tunnelling microscopy and EELS in transmission in which chiral as well as non-chiral ('armchair' and 'zig-zag') nanotubes were positively identified [8,15]. Additionally, it has been shown that bundles of SWNTs can be intercalated with alkali metals or halogens in order to achieve n- or p-type doping, respectively [16,17].

In this contribution we present high resolution momentum-dependent EELS in transmission measurements of

---

*Corresponding author.

purified SWNTs, in which we show that they support two types of electronic excitations. The first group of excitations are non-dispersive, whereby their energy position is characteristic of the separation of the van Hove singularities in the electronic density of states of the different types of nanotubes present in the sample. The second type of excitation shows considerable dispersion, which parallels that observed in the graphite plane, and is related to a collective excitation of the π-electron system polarized along the nanotube. In addition, the close similarity to graphite is also indicated by the C1s excitation edges.

2. Experimental

SWNTs were produced by a laser vaporization technique [18]. The material consists of up to 60% SWNTs with approximately 1.4 nm mean diameter and was purified as described in Ref. [19]. Free-standing films for EELS of effective thickness about 1000 Å were prepared on standard copper microscopy grids via vacuum filtration of a nanotube suspension in a 0.5% surfactant (Triton X100) solution in de-ionised water, with a SWNT concentration of ~0.01 mg ml^{-1}. The surfactant was them rinsed off and the film was transferred into the spectrometer. EELS was carried out in a purpose-built high-resolution spectrometer [20] which combines both good energy *and* momentum resolution. For the data shown here, an energy and momentum resolution of 115 meV and

0.05 Å^{-1} were chosen. Unlike many electron spectroscopy techniques, this method has the advantage of being volume sensitive. In addition, by setting the energy-loss to zero, we are able to characterize the films *in-situ* using electron diffraction. In Fig. 1 we show the electron diffraction profile of thin SWNT films used in our studies. The spectrum is consistent with the triangular SWNT lattice in the bundles with the peak at about 0.42 Å^{-1} representing the nearest neighbour distance of the SWNTs of about 15 Å. The spectrum shown in Fig. 1 is in agreement with X-ray diffraction results [18] and demonstrates the clean and crystalline nature of the SWNT films studied here.

3. Results and discussion

In Fig. 2 we present the EELS C1s excitation spectrum of purified SWNT compared to that of highly oriented pyrolytic graphite (HOPG, measured with the momentum transfer perpendicular to the graphite plane). In such a measurement, excitations of the core electron into unoccupied states with C2p character are probed. The spectrum of the SWNTs strongly resembles that of graphite. By comparison with directional-dependent C1s excitation measurements of HOPG [21,22], we see that the SWNT spectrum most closely resembles an average of the in-plane and out-of-plane graphite spectra. The predicted singularities in the unoccupied C2p-derived electronic states of the SWNTs [4–6] are conspicuous by their

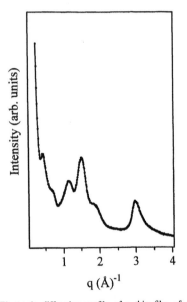

Fig. 1. Electronic diffraction profile of a thin film of purified SWNT material.

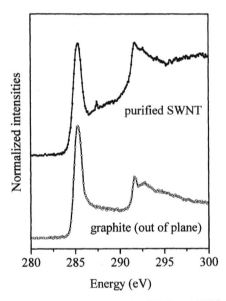

Fig. 2. C1s excitation spectra of purified SWNTs and HOPG (out of plane).

absence in Fig. 2. We attribute this fact to the effect of the C1s core hole in the final state. In the case of C1s excitation spectroscopy of graphite, both the π^* [23–26] and σ^* [22,26] onsets are dominated by spectral weight resulting from the influence of the core hole. Thus, assuming a similar interaction between the excited electron and the core-hole in SWNTs as in graphite, we can expect that the π^* resonances related to the density of states (DOS) singularities of the different types of SWNTs are washed out, resulting in the broad peak centred at 285.2 eV. Intriguingly, a small, sharp feature is observed in the pristine SWNT spectrum at 287.4 eV. There is no analogue of this feature in the spectra of HOPG and its origin remains unclear at this point.

In Fig. 3 we show the loss function $(\text{Im}(-1/\epsilon(q, E)))$ for the purified SWNTs measured as a function of q, after the subtraction of the quasi-elastic line [27]. The inset

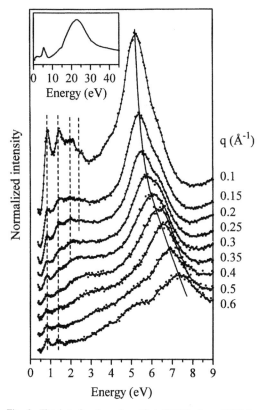

Fig. 3. The loss function of purified SWNTs from EELS in transmission for the different q values shown. The contributions from the elastic peak have been subtracted. The inset contains the loss function over an extended energy range for $q=0.15$ Å^{-1}, showing the π-plasmon and the $\pi+\sigma$-plasmon at around 5 and 22 eV, respectively.

shows the loss function for $q=0.15$ Å^{-1} over a wide energy range, in which the π plasmon, which represents the collective excitation of the π-electron system, can be clearly seen at an energy of 5.2 eV, and the $\sigma+\pi$ plasmon (the collective excitation of all valence electrons) at 21.5 eV. These values for SWNTs are in agreement with the spatially-resolved EELS data mentioned above [12] and confirm theoretical predictions that the π-plasmon should occur in the energy range of 5–7 eV in the EELS of these materials [28–30]. We note that the observation of the π-plasmon at its characteristic energy position requires the arrangement of the SWNTs in bundles as otherwise this *collective* excitation would not be possible. Fig. 3 shows that we are able to reliably measure the q-dependence of excitations in SWNTs with energies as low as 0.5 eV, which will be shown in the following to be vital to the understanding of their dielectric response. Furthermore, these data provide a wealth of information not accessible in spatially-resolved EELS measurements, in which the study of momentum-dependence is excluded and the broad quasi-elastic tail of the direct beam has only allowed the extraction of reliable information for energies above ~3 eV [9–12].

The loss function measured in an EELS experiment is a direct probe of the collective excitations of the system under consideration. Thus, by definition, all peaks in the measured loss function should be considered as plasmons. These peaks can, however, have different origins such as charge carrier plasmons, interband or intraband excitations. From the q-dependence of the loss function, one can distinguish directly between features arising from localized or delocalized electronic states. Localized states give rise to a vanishingly small dispersion of peaks in the loss function as has been observed, for example, for the features related to the interband excitations and both the π and $\pi+\sigma$ plasmons of C_{60} [31]. On the other hand, excitations between delocalized states generally exhibit a band structure dependent dispersion relation. Bearing these points in mind, the identification of excitations between localized and delocalized states in SWNTs is straightforward. At low momentum transfer, features in the loss function are visible at about 0.85, 1.45, 2.0, 2.55, 3.7, 5.2 and 6.4 eV whose origin lies in the π electron system of the SWNTs [32]. However, two distinct behaviours of these features as a function of q are observed. The π plasmon disperses strongly from 5.2 eV at $q=0.1$ Å to 7.4 eV at $q=0.6$ Å^{-1}, whereas all the other peaks have a vanishingly small dispersion.

The momentum dependence of the peaks in the loss function resulting from the four lowest lying interband transitions of the SWNTs, as well as those of the π and $\pi+\sigma$ plasmons is summarized in Fig. 4. For comparison we also show the dispersion of the π and $\pi+\sigma$ plasmons of graphite with q parallel to the planes. In low dimensional systems, the nature of the plasmon excitations depends on their polarization. This has been shown, for example,

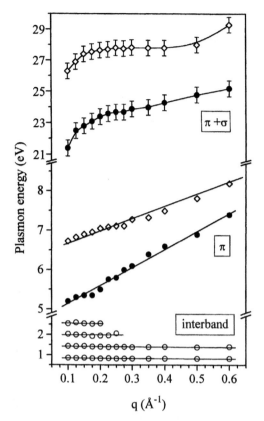

Fig. 4. The dispersion of the π, $\pi+\sigma$ plasmons ($\bullet$), and of the features arising from interband transitions between localized states (O) in purified SWNTs from EELS in transmission measurements. When invisible, the error bars are within the size of the symbols. For comparison the dispersion of the π and $\pi+\sigma$ plasmons in graphite for momentum transfer parallel to the planes is also shown as ($\diamond$).

for oriented trans-polyacetylene, whereby a dispersive plasmon is only visible in the one-dimensional direction [33].

Thus, in combination with the well-known one-dimensionality of nanotubes the nondispersive peaks in the loss function can be attributed to excitations between localized states polarized *perpendicular* to the nanotube axis and thus resemble molecular interband transitions such as those of C_{60}. In contrast, the π-plasmon (at 5.2 eV for low $\mathbf{q}$), represents a plasma oscillation of delocalized states polarized *along* the nanotube axis. As can be seen from Fig. 4, the dispersion relations of both the π and the $\pi+\sigma$ plasmons in SWNTs and graphite are very similar, which confirms the graphitic nature of the axial electron-system in carbon nanotubes.

By carrying out a Kramer–Kronig analysis (KNA) of

the loss function we can derive the real (ϵ_1) and imaginary parts (ϵ_2) of the dielectric function. The results of such a KKA are depicted in the upper panels of Fig. 5 for C_{60}, SWNTs, and graphite (measured in the plane). In the lowest panel the corresponding real part of the optical conductivity is plotted, whereby $\sigma_r(E)=(E/\hbar)\epsilon_0\epsilon_2$ is a measure of the joint density of states. For the KKA of the SWNTs the loss function was normalized using an estimated $\epsilon_r(q, 0)$ [34]-a procedure which has been used successfully in the past for graphite [35,36].

In general, the optical conductivity of these sp^2 conjugated carbon systems show peaks due to transitions between the (π/σ) and the (π^*/σ^*) electronic states. In C_{60} these peaks are very pronounced which is consistent with the high symmetry of the molecule and the weak, van der Waals interactions in the solid state [37], making C_{60} a prototypical zero-dimensional solid. In graphite three broad features are observed at 4.5 ± 0.05, 13 ± 0.05, and 15 ± 0.05 eV. Their breadth is an expression of the bandlike nature of the electronic states in the graphite plane.

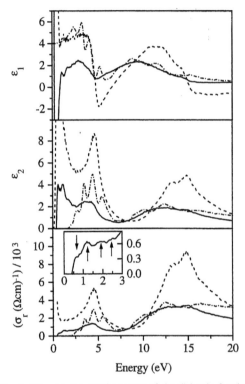

Fig. 5. The real and imaginary part of the dielectric function (upper panels) and the real part of the optical conductivity (σ_r) at low momentum transfer: SWNTs (solid line) at $q=0.1$ Å^{-1}, C_{60} (dotted–dashed line) and graphite (in plane, (dashed line)) at $q=0.15$ Å^{-1}, respectively. The inset shows σ_r for the four lowest lying interband transitions of SWNTs in an expanded range.

For the SWNTs we also find three broad features at energies slightly lower than those in graphite-i.e. 4.3 ± 0.1, 11.7 ± 0.2, and $14.6+0.1$ eV. Importantly, the optical conductivity of the SWNTs also exhibits additional structures at low energy. This region is shown in detail in the inset of Fig. 5, where three pronounced interband transitions are seen at energies of 0.65 ± 0.05, 1.2 ± 0.1 and 1.8 ± 0.1 eV. Three further features, which are less pronounced, are located at 2.4 ± 0.2, 3.1 ± 0.2 and 6.2 ± 0.1 eV.

As the SWNT data in the lowest panel (and inset) of Fig. 5 represent joint densities of states, we can directly relate the energy position of the features with the energetic separation of the one-dimensional van Hove singularities in the electronic density of states. It is known from X-ray diffraction that these samples of SWNTs have a narrow diameter distribution around a mean value of 1.4 nm [18]. The fact that we observe well defined non-dispersive features in the EELS data confirms a narrow diameter distribution, as otherwise the sheer number of energetically different interband transitions would wash out all fine structure in this energy range, both in the loss function or in the optical conductivity.

The origin of the features at lowest energy (0.65 and 1.2 eV) is unambiguous-for the nanotube diameter range relevant here, only the gap transition and the transition between the next pair of DOS singularities in semiconducting nanotubes are predicted to lie at such energies [1–6]. STS experiments on single nanotubes, which were characterized using STM, have recently confirmed these predictions [8]. The peak appearing in the optical conductivity at 1.8 eV, corresponds directly to the "gap" between the DOS singularities straddling E_F which have been observed experimentally in STS of metallic chiral and zigzag tubes [8]. Thus the feature at 1.8 eV in the optical conductivity clearly originates from metallic nanotubes. For the features at higher energies, an assignment based upon a discussion of particular nanotubes becomes less secure due to the large number of possible optically allowed transitions.

4. Summary

In summary, we have demonstrated that momentum-dependent high-resolution EELS in transmission measurements of SWNTs show that they represent a text-book example of a system supporting both excitations between localized and delocalized electronic states. Along the nanotube axis SWNTs support a π-plasmon whose dispersion relation is very similar to that of the graphite plane, proving the graphitic nature of the electronic system in this direction. At low energies, non-dispersive features are observed whose energy location in the real part of the optical conductivity is directly related to the energy separation of the DOS singularities in the nanotubes. Thus, assuming the matrix elements governing the EELS transitions for different nanotubes to be comparable, we prove

the presence of a significant proportion of semiconducting nanotubes in these bulk samples in addition to the metallic nanotubes inferred from other measurements [14,18]. These data show that EELS in transmission offers a source of information regarding the character (semiconducting/metallic) of carbon nanotubes in *bulk, macroscopic* samples-which makes it an ideal complement to spatially-resolved measurements of individual SWNTs. Moreover, the differences in q-dependence of the features polarized along and perpendicular to the nanotube axis make momentum-dependent EELS an ideal probe of anisotropies in oriented bulk samples of SWNTs in the future.

Acknowledgements

T.P. thanks the European Union for funding under the 'Training and Mobility of Researchers' program. The work at Rice was supported by the National Science Foundation (DMR9522251), the Advanced Technology Program of Texas (003604-047) and the Welch Foundation (C-0689).

References

[1] Saito S. Science 1997;278:77.
[2] Chico L, Crespi VH, Benedict LX, Louie SG, Cohen ML. Phys Rev Lett 1996;76:971.
[3] Wong EW, Sheehan PE, Lieber CM. Science 1997;277:1971.
[4] Mimitmire JW, Dunlop BI, White CT. Phys Rev Lett 1992;68:631.
[5] Hamada N, Sawada SI, Oshiyama A. Phys Rev Lett 1992;68:1579.
[6] Dresselhaus MS, Dresselhaus G, Eklund PC. Science of Fullerenes and Carbon Nanotubes. San Diego: Academic Press, 1996.
[7] Tans SJ, Devoret MH, Dai HJ, Thess A, Smalley RE, et al. Nature 1997;386:474.
[8] Wildör JWG, Venema LC, Rinzler AG, Smalley RE, Dekker C. Nature 1998;391:59.
[9] Kuzuo R, Terauchi M, Tanaka M. Jpn J Appl Phys 1992;31:L1484.
[10] Ajayan PM, Ijima S, Ichihashi T. Phys Rev B 1994;49:2882.
[11] Bursill LA, Stadelmann PA, Peng JL, Prawer S. Phys Rev B 1994;49:2882.
[12] Kuzuo R, Terauchi M, Tanaka M, Saito Y. Jpn J Appl Phys 1994;33:L1316.
[13] Petit P, Jouguelet E, Fischer JE, Rinzler AG, Smalley RE. Phys Rev B 1997;56:9275.
[14] Rao AM, Richter E, Bandon S, Chase B, Eklund PC, et al. Science 1997;275:187.
[15] Pichler T, Knupfer M, Golden MS, Fink J, Rinzler A, et al. Phys Rev Lett 1998;80:4729.
[16] Rao AM, Ekland S, Bandon S, Thess A, Smalley RE. Nature 1997;388:257.
[17] Lee RS, Kim HJ, Fischer JE, Thess A, Smalley RE. Nature 1997;388:255.
[18] Thess A, Lee R, Nikolaev P, Dai H, Petit A, et al. Science 1996;273:483.

[19] Rinzler AG, Liu J, Dai H, Nikolaev P, Huffman CB, et al. Appl Phys A 1998;67:29.

[20] Fink J. Adv Electron Electron Phys 1989;75:121.

[21] Batson PE. Phys Rev B 1993;48:2608.

[22] Ma Y, Skytt P, Wassdahl N, Glans P, Mariani DC, et al. Phys Rev Lett 1993;71:3725.

[23] Mele EJ, Ritsko JJ. Phys Rev Lett 1979;43:68.

[24] Brühwiler P, Maxwell AJ, Puglia C, Nilsson A, Anderson S, et al. Phys Rev Lett 1995;74:614.

[25] van Veenendaal M, Carra P. Phys Rev Lett 1997;78:2839.

[26] Shirley EL. Phys Rev Lett 1998;80:794.

[27] Golden MS, Pichler T, Knupfer M, Fink J, Rinzler AG, Smalley RE. In: Kuzmany H, Roth S, Fink J, Mehring M, editors. Proceedings of the International Winterschooh on Electronic Properties of Novel Materials 1998, Kirchberg, Tirol, 1998, in press.

[28] Lucas AA, Henrad L, Lambin Ph. Phys Rev B 1994;49:2888.

[29] Lin MF, Shiung KWK. Phys Rev B 1994;50:17744.

[30] Lin MF, Chuu DS, Huang CS, Lin YK, Shung KWK. Phys Rev B 1996;53:15493.

[31] Romberg H, Sohmen E, Merkel M, Knupfer M, Alexander M, et al. Synthetic Metals 1993;55–57:3038.

[32] As is often the case in EELS at high momentum transfers, contributions from the $q=0$ spectrum re-appear due to quasi-elastic scattering. This is the case here for $q \geq 0.5$ Å^{-1}.

[33] Fink J, Leising G. Phys Rev B 1986;34:5320.

[34] Venghaus H. Phys Stat Sol 1974;66:154. Due to the presence of the quasi-elastic line at extremely low energies ($E<0.4$ eV) EELS is unable to deliver directly an experimental value for $\epsilon_r(0)$. For the SWNTs we have taken $\epsilon_r(0)$ to be 100 ($q=0.1$ Å^{-1}), which is arrived at from an average of the values for graphite in the plane [35] and perpendicular to the plane. It is important to note, however, that the energy positions of the interband transitions in the optical conductivity remain essentially unchanged if a metallic extrapolation to zero energy is assumed (data not shown).

[35] Zeppenfeld K. Z Phys 1971;243:229.

[36] Ritsko JJ, Mele EJ. Phys Rev B 1980;21:730.

[37] Sohmen E, Fink J. Phys Rev B 1993;47(14):532.

Pergamon

Carbon 37 (1999) 739–744

CARBON

Scanning probe method investigation of carbon nanotubes produced by high energy ion irradiation of graphite

L.P. Biró[a,d,*], G.I. Márk[a], J. Gyulai[a,c], N. Rozlosnik[b], J. Kürti[b], B. Szabó[a,b], L. Frey[c], H. Ryssel[c]

[a]Research Institute for Technical Physics and Materials Science, H-1525 Budapest, P.O. Box 49, Hungary
[b]Eötvös University, Department of Biological Physics, H-1088 Puskin u.5-7, Hungary
[c]FhG-Institut für Integrierte Schaltungen, 91058 Erlangen, Schottkystr 10, Germany
[d]Facultes Universitaires Notre Dame de la Paix, LASMOS, 5000 Namur, Rue de Bruxelles 61, Belgium

Received 16 June 1998; accepted 3 October 1998

Abstract

Carbon nanotubes were evidenced by atomic force microscopy and scanning tunneling microscopy on highly oriented pyrolytic graphite irradiated with high energy ions (215 MeV Ne, 209 MeV Kr, 246 MeV Kr, and 156 MeV Xe). On the samples irradiated with Kr and Xe ions, craters attributed to sputtering were found. Frequently, one or several nanotubes emerge from these sputtering craters. Some of the observed nanotubes vibrate when scanned with the AFM. Except nanotubes, no other deposits were observed. © 1999 Elsevier Science Ltd. All rights reserved.

Keywords: A. Carbon nanotubes; B. Plasma sputtering; C. Atomic force microscopy; Scanning tunneling microscopy

1. Introduction

Since their discovery in 1991 by Iijima [1], due to their remarkable properties [2,3], a continuously growing interest is focused onto the carbon nanotubes (CNT). Recent experimental results proved an excellent agreement between theoretical predictions [4] and the electronic structure of single-walled carbon nanotubes (SWNT) probed by scanning tunneling spectroscopy (STS) [5,6]; experimental evidence pointing to the feasibility of CNT based nanoelectronics were reported [7,8].

The most frequently used procedures to produce large amounts of carbon nanotubes are: the electric arc method [9], laser ablation [10], and the catalytic decomposition of hydrocarbons [11]. All these methods yield amorphous carbon and graphitic material together with the CNT. Tedious purification steps are needed to obtain a material which contains only CNTs. In the electric arc, and in the laser ablation methods, the growth of carbon nanotubes takes place in a highly excited carbon plasma generated at several thousands °C, while the catalytic process is based on the dehydrogenation at 700°C of hydrocarbons. In the present paper we report a new procedure for the production of carbon nanotubes which is based on the irradiation of highly oriented pyrolytic graphite (HOPG) targets with high energy ($E > 100$ MeV) heavy ions. Individual CNTs of several μm in length are observed by atomic force microscopy (AFM) and scanning tunneling microscopy (STM) in, or around the craters where higher order, dense nuclear cascades – the so called Brinkman type cascades [12] – intersected the sample surface and produced extended sputtering. Frequently, the CNTs traversing surface features on the HOPG vibrate when scanned with a tapping mode (TM) AFM.

2. Experimental results and discussion

The HOPG samples were irradiated with low dose, typically 10^{12} cm^{-2}, of 215 MeV Ne, 209 MeV and 246 MeV Kr and 10^{11} cm^{-2} 165 MeV Xe ions. The samples were freshly cleaved before irradiation and were handled with maximum care after irradiation to avoid any surface contamination. It is worth mentioning that in the last years

---

*Corresponding author. Tel: +36-1-3959220; fax: +36-1-3959284.
E-mail address: biro@mfa.kfki.hu (L.P. Biro)

we investigated in some detail the interaction of swift, heavy ions with crystalline targets, like: Si [13,14], muscovite mica [15], and HOPG [16,17]. Usually, Si, mica and HOPG samples were irradiated simultaneously, during the same run. While no CNTs were found on any of the Si or mica samples, on every HOPG sample we found CNTs.

The irradiated samples were investigated in ambient atmosphere using contact mode (CM) and tapping mode (TM) atomic force microscopy, and scanning tunneling microscopy. For AFM measurements blunt tips ($R > 100$ nm) were preferred because the convolution effect of the tip shape with the tube shape produces the apparent widening of the tube in the horizontal plane. This facilitates finding the CNTs in large scan windows of the size of 100 μm^2. The STM tips were mechanically prepared Pt tips, the tunneling parameters used were similar to those commonly used for HOPG: tunneling current of 1 nA, and bias of 100 mV.

In Fig. 1 a TM-AFM image of a sputtering crater on the surface of HOPG irradiated with Xe ions is shown, the total length of the carbon nanotube emerging from the crater is 11 μm, the diameter of the CNT as measured from its height – the height is not influenced by convolution effects – is 6 nm in the vicinity of the crater and gradually decreases to 1 nm in the end region. As shown by the line cut in Fig. 1, the tube height as measured along the tube axis shows a regular oscillation pattern. It is worth pointing out that the edge of the crater does not show any oscillatory behavior, i.e., the oscillation of the CNT is not an imaging artifact. Assuming a conic shape for the crater, its estimated volume is 1.4×10^{-3} μm^3, this means that 1.6×10^8 carbon atoms were sputtered away. Taking the CNT in Fig. 1 as being a double wall tube, and assuming that all the atoms will be built into CNTs, 20 similar CNTs could be generated from the C atoms sputtered out of the crater, if the density of bulk graphite is used, this number decreases to 14.

Those sputtered atoms which have a significantly higher energy will quickly leave the vicinity of the surface, while the less energetic ones, and the clusters, will form an expanding cloud in which the nanotube growth may take place. After the fast growth in the expanding cloud takes place, the nanotube will collapse onto the surface, where additional growth may take place due to the target atoms ejected at very low energies in those collisions which are responsible for the generation of the well known hillocks on ion bombarded HOPG [18]. For example, in the case of 246 MeV Kr irradiation it is found by STM examination that 10^2 μm^{-2} hillocks are produced, while the density of carbon nanotubes found is 2×10^{-3} μm^{-2}. For a dose ten times lower, in the case of Xe irradiation a slightly higher nanotube density in the range of 3×10^{-3} μm^{-2} was found.

It was frequently observed that especially those carbon nanotubes, which cross elevated surface features, like cleavage steps or folds on HOPG, oscillate when scanned

with the AFM. A TM-AFM image showing this phenomenon is presented in Fig. 2, the image was not filtered in any way. One can note that the step crossing the image does not show any vibration, the nanotube crosses the step close to the left hand edge of the image. The total length of the nanotube exceeds 100 μm, its diameter is estimated to be 36 nm, the vibration causes some incertitude in the accuracy of this value.

To get a better insight into how the TM-AFM generates the image of a vibrating object, computer modeling was used. As a first approximation a simple model consisting of a cylindrical rod of 36 nm diameter performing a transversal lateral sinusoidal vibration was used. The rod is placed on a flat support surface along the y direction. The scan direction was perpendicular to the rod axis. The scan was performed in a window of 50×50 μm^2 at a scan frequency of 1 Hz by a tip of spherical shape with a radius of 300 nm. The window was sampled in 256×256 pixels, with an averaging of 64 kHz, the sampling frequency used in the TM-AFM. We have assumed that in each sampling point the tip is descending until it touches the sample. From this assumption for a cylindrical rod of radius R_{rod} at position x_{rod} and a spherical tip of radius R_{tip} at lateral position (x_{tip}, y_{tip}) the apparent tip height above the rod is obtained as:

$$z_{tip} = [(R_{rod} + R_{tip})^2 - (x_{rod} - x_{tip})^2]^{(1/2)} - R_{rod}, \quad (1)$$

where

$$x_{rod} = A \sin(2\pi f t), \quad (2)$$

and x_{tip} is increasing linearly in each scan line with a scan speed of $v_{tip} = 100$ $\mu m/s$.

From this simple model similar vibration patterns like those measured experimentally were obtained, a 3D presentation comparable with Fig. 2 is shown in Fig. 3. To make a semiquantitative comparison we used the spatial period of 5 μm and corrugation amplitude of 15 nm determined from an experimental line cut drawn along the tube axis. From these parameters we have obtained $f = 1/51.2$ Hz and $A = 96.5$ nm for the vibration frequency and amplitude of the rod. The surprisingly low vibration frequency may be a consequence of the 'friction' of the nanotube with the supporting surface [19,20]. Further work is needed to built models that take in account these interactions and simulate in a more realistic way the dissipation of the vibration energy fed in the cantilever-tip system.

We finally consider a possible scenario for the crater formation and material sputtering. There are two mechanisms by which an energetic particle moving in a solid may transfer enough energy to the solid to induce structural modifications: (i) the electronic stopping mechanism (inelastic interaction with the electrons of the target atoms), typical for high energy particles ($E > 1$ MeV); and (ii) the nuclear stopping mechanism (elastic collisions with

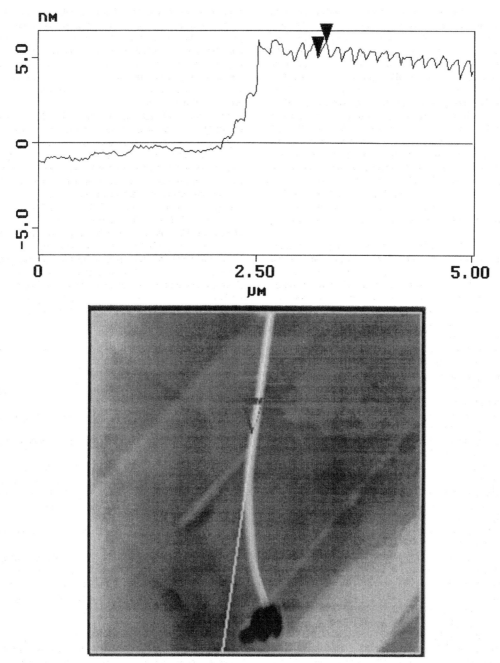

Fig. 1. Tapping mode AFM image of a sputtering crater and of a nanotube emerging from the crater on the surface of HOPG irradiated with Xe ions. The line cut over the image shows the presence of a regular oscillation pattern along the tube axis. The vertical amplitude of the oscillation is 0.7 nm, while its spatial period is 194 nm.

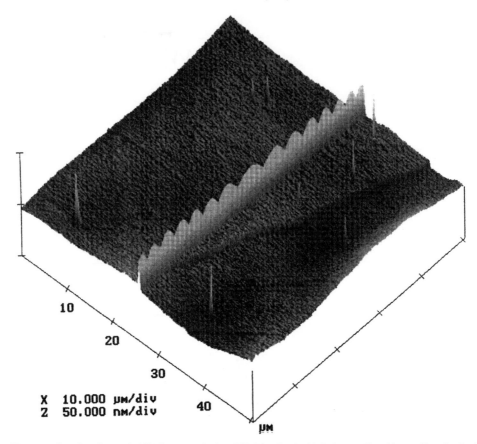

X 10.000 µm/div
Z 50.000 nm/div

Fig. 2. 3D presentation of tapping mode AFM image acquired on HOPG irradiated with Kr ions, unfiltered image. Note the simultaneous presence of a step which is immobile in the image, and of a carbon nanotube, which shows a regular oscillation pattern. The diameter of the nanotube is estimated to be 36 nm.

the nuclei of the target atoms) dominant for low energy particles. The high energy ions used in the present work, when falling on the target surface, they all have energies in the dominant electronic stopping regime. So, if this mechanism would be responsible for the crater production, every incident ion should generate a crater. The surface densities of the craters when Kr or Xe ions were used, were in the range of 10^{-3} μm^{-2}, this means that for example in the case of Kr ions every 10^7 ions produces an event that leads to a surface crater. Therefore, the crater production and surface sputtering is attributed to those knocked-on target atoms which are produced by elastic collisions in deeper regions of the target, in higher order cascades. These target atoms are displaced by other target atoms which have gained kinetic energy in earlier collisional processes. Some of the C atoms knocked-on in the higher order collisions may take an outward pointing direction, intersecting the irradiated surface. If such a knocked-on atom generates in turn several sub-cascades in

a limited region, which are active in the same time, this may lead to extended surface sputtering in that region where the simultaneously active sub-cascades cross the target surface [21]. In an earlier work we reported TM-AFM images of the surface termination of dense nuclear cascades in Si and muscovite mica [22], in which numerous satellites are clearly visible around the main cascade. Due to the particular structure of HOPG this kind of cascades may produce the observed craters. Although, at the present moment the above detailed scenario seems to be the most realistic one that accounts for the crater production, some synergistic interaction of electronic and nuclear stopping effects may not be excluded.

In order to get additional proof that the observed objects are different from surface folds on HOPG, some of the nanotubes and an object, clearly identified as a surface fold, were cut by a focused ion beam (FIB) apparatus and the section was examined by the scanning electron microscope built into the FIB. When comparing cut nanotubes

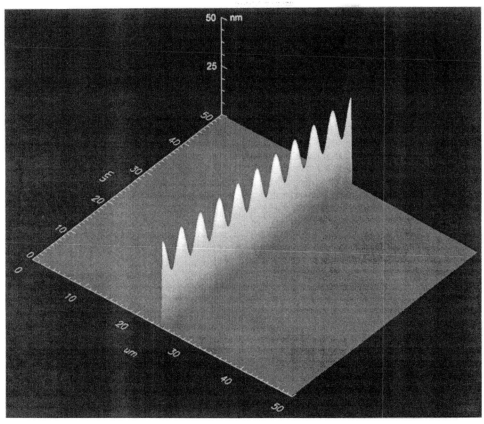

Fig. 3. Computer simulation of the image of a vibrating rod in the tapping mode AFM. An amplitude of 15 nm and a spatial period of 5 μm were chosen for the rod for easier comparison with the experimental image shown in Fig. 2.

with the fold, clear differences were found in the cross-sectional structure of these objects [21].

3. Conclusions

Carbon nanotubes of several μm in length, produced from target atoms sputtered by dense, nuclear cascades were evidenced by AFM and STM investigation of HOPG targets irradiated with high energy, heavy ions. The formation of other carbon based structures, like graphitic particles or amorphous carbon was not observed. This is an indication that using ion beam techniques carbon nanotubes may be produced in a way which makes possible the avoidance of purification steps.

Frequently, one or several nanotubes were found to emerge from the sputtering craters. Several vibrating nanotubes were observed, a comparison with a simple computer model of the image generation of a vibrating rod in a TM-AFM indicates an oscillation frequency of the order of 0.02 Hz. The extremely low frequency is tenta-tively attributed to the interaction of the nanotube with the support.

Acknowledgements

Helpful discussions with Prof. Ph. Lambin and Prof. P.A. Thiry of FUNDP Namur, are gratefully acknowledged. This work was supported in Hungary by OTKA T025928, AKP 96/2-637 and AKP 96/2-462 grants. L.P. Biró gratefully acknowledges a fellowship from the Belgian SSTC for S and T cooperation with Central and Eastern Europe.

References

[1] Iijima S. Nature 1991;354:56.
[2] Dresselhaus MS, Dresselhaus G, Eklund PC. Science of fullerenes and carbon nanotubes. San Diego: Academic Press, 1996.

[3] Ebbesen ThW, editor. Carbon nanotubes preparation and properties. Boca Raton: CRC Press, 1997.

[4] Mintmire JW, Dunlap BI, White CT. Phys Rev Lett 1992;68:631.

[5] Wildöer JWG, Venema LC, Rinzler AG, Smalley RE, Dekker C. Nature 1998;391:59.

[6] Odom TW, Huang J-L, Kim Ph, Lieber ChM. Nature 1998;391:62.

[7] Collins PhG, Zettl A, Brando H, Thess A, Smalley RE. Science 1997;278:100.

[8] Tans SJ, Verschueren ARM, Dekker C. Nature 1998;393:49.

[9] Ebbesen TW, Hiura H, Fujita J, Ochiai Y, Matsui S, Tanigaki K. Chem Phys Lett 1993;209:83.

[10] Thess A, Lee R, Nikolaev P, Dai H, Petit P, Robert J, Xu C, Lee YH, Kim SG, Rinzler AG, Colbert DT, Scuseria GE, Tománek D, Fischer JE, Smalley RE. Science 1996;273:483.

[11] Ivanov V, B Nagy J, Lambin Ph, Lucas A, Zhang XB, Zhang XF, Bernaerts D, Van Tendeloo G, Amellinckx S, Van Landuyt J. Chem Phys Lett 1994;223:329.

[12] Brinkman AJ. Am J Phys 1956;25:961.

[13] Biró LP, Gyulai J, Havancsák K, Didyk AYu, Bogen S, Frey L. Phys Rev 1996;B 54:11853.

[14] Biró LP, Gyulai J, Havancsák K, Didyk AYu, Frey L, Ryssel H. Nucl Instr Meth B 1997;127–128:32.

[15] Biró LP, Gyulai J, Havancsak K. Nucl Instr Meth 1997;B 122:476.

[16] Biró LP, Gyulai J, Havancsák K. Phys Rev 1995;B 52:2047.

[17] Ziegler JP, Biersack JP, Littmark V. The stopping and range of ions in solids I and II. New York: Pergamon Press, 1986.

[18] Coratger R, Claverie A, Ajustron F, Beauwillain J. Surf Sci 1990;227:7.

[19] Falvo MR, Clary GJ, Taylor RM, Chi V, Brooks Jr. FP, Washburn S, Superfine R. Nature 1997;389:582.

[20] Wong EW, Sheenan PE, Lieber ChM. Science 1997;277:1971.

[21] Biró LP, Márk GI, Gyulai J, Havancsák K, Lipp S, Lehrer Ch, Frey L, Ryssel H. Nucl. Instr. Meth. B: in press.

[22] Biró LP, Gyulai J, Havancsák K. Vacuum 1998;50:263.

Pergamon

Carbon 37 (1999) 745–752

CARBON

Field emission from diamond, diamond-like and nanostructured carbon films

Olivier M. Küttel*, O. Gröning, Ch. Emmenegger, L. Nilsson, E. Maillard, L. Diederich, L. Schlapbach

University of Fribourg, Physics Department, Pérolles, 1700 Fribourg, Switzerland

Received 16 June 1998; accepted 3 October 1998

Abstract

We have deposited nanotube films on silicon via a chemical vapor deposition (CVD) growth process known from the deposition of diamond. We used a metallic catalyst which was deposited onto the silicon surface prior to the CVD deposition. The films are very pure, adhere well and are very well suited for electron field emission. We measured emission at 2.6 V/μm (for 1 nA emission current) and an emission site density reaching 10^4/cm^2 at 3–4 V/μm as measured on a phosphor screen. Electrons originate at the Fermi level and the high local fields at the emission site is produced by the geometry of the nanotube. The results obtained on these films are comparable to those from differently prepared CVD diamond films. So far, we have no evidence that electron injection occurs. The emission process is governed by field amplification at protrusions and tips. In a second experiment we have measured emission from a metallic micrometer sized grain fixed on a diamond (100) surface, with different surface termination (hydrogen, oxygen, sp^2 carbon). The field emitted electron energy distribution (FEED) spectra show large energy shifts which are due to the surface resistivity and not due to injection of electrons in the conduction band. Hence, energy shifts in FEED spectra do not necessarily reflect an injection mechanism. © 1999 Elsevier Science Ltd. All rights reserved.

Keywords: A. Carbon films; Diamond; Diamond-like carbon; B. Chemical vapor deposition; D. Electronic structure; Field emission

1. Introduction

Field emission from carbon films is widely recognized as being extraordinary in many ways. First of all electron emission is observed for modest applied macroscopic fields (1–20 V/μm), and secondly, it is possible to get electron emission from flat surfaces. Such films are very promising candidates for applications as flat panel displays (FPD), cold cathodes and others. Flat panels require an emission site density of approx. 1–10 million spots per cm^2 in order to have a reasonably good statistic which smoothes out the current fluctuations of individual emitters. Carbon films emitting electrons at electric fields lower than 10 V/μm are very well documented in the literature [1–5]. Many different deposition techniques can be used including filtered arc deposition, CVD diamond growth, laser ablation, ECR plasma deposition to mention just a few.

So far, different emission models are discussed [1,6–11] however, none have yet proven to fully explain the observed results. It might be that different mechanisms are involved. Discussed models are injection into the conduction band, field focusing which we will refer to as Latham's model, [7] and classical field emission from carbon protrusions and tips.

In the past, we have shown that electron emission can be activated [3]. Increasing the applied electric field on a freshly deposited sample in many cases leads to a very short arc between the anode and the cathode. During this vacuum discharge the surface morphology of the film is changed, very likely tips and protrusions are formed which in turn leads to a pronounced emission. Even though this mechanism was discussed by different authors [8,12,13], many groups still are not aware of it. It is known from investigations on high voltage insulation that conditioning is a wide spread phenomenon which is not limited to carbon. Any metal surface shows such a behavior. For high voltage insulation, electron field emission has to be

---

*Corresponding author.
E-mail address: olivier.kuettel@unifr.ch (O.M. Kuettel)

0008-6223/99/$ – see front matter © 1999 Elsevier Science Ltd. All rights reserved.
PII: S0008-6223(98)00265-6

avoided. However, it is known that field emission occurs at fields which are much lower than what one would expect. Several phenomena are observed when increasing the voltage between two electrodes. The macroscopic field is locally enhanced by imperfections or protrusions leading eventually to electron emission. As the field is increased further a spark occurs. Such a spark is characterized by high current densities which leads to morphological changes at the surfaces. We have shown that in some cases these discharges can be very violent. Many times, it is nearly impossible to detect these discharges except from measuring a higher emitted electron current. Such a behavior is not typical for carbon only, but was observed on other electrodes. It is important to note that a conditioning or activation process leads to some kind of changes at the surface of the investigated film. Hence, when performing a conditioning process like ramping the voltage, maintaining the voltage at a certain level or some other processes, the film undergoes changes. Films using a conditioning step are not really useful, especially when trying to correlate the film structure with the emission properties. Another phenomenon which is rarely addressed when dealing with emission from carbon films is gas desorption from unannealed electrodes. It is known from investigations on metal electrodes that gas desorption is driven by anode conditioning as the voltage is increased. The gas is often desorbed as microdischarges, leading to some kind of morphological changes. Microdischarges can then trigger field-emitted electrons which, in turn, heats a small area of the anode.

Very often electron field emission (EFE) is characterized by measuring I/V curves and computing the so-called Fowler-Nordheim (FN) plot. Form the slope of such a plot one can get information about the work function ϕ and the field enhancement factor β. However, it is not possible to independently determine both values. Hence, the work function which is computed from a FN plot depends very much on the β factor. Many times a β factor of unity is used in literature what is certainly false, as carbon tends to form protrusions, clusters and tips. On the other hand it is very difficult to experimentally determine curvatures and β factors for example by high resolution scanning electron microscopy (HRSEM). To our knowledge there is only one method which at the same time independently gives ϕ and β the energy resolved measurement of the field emitted electron distribution (FEED) [14–16]. From the width of the peak in the FEED spectrum and a simultaneously performed I/V plot one gets two pieces of information to compute ϕ and β.

In Fig. 1 we have plotted the relation between ϕ and β as measured from a CVD grown nanoclustered diamond film on molybdenum. The dashed line reflects the FEED measurements and the solid line represents what one gets from the I/V plot. The I/V plot is realized by measuring several FEED spectra for different applied potentials. By integrating the FEED peaks the emitted current can be

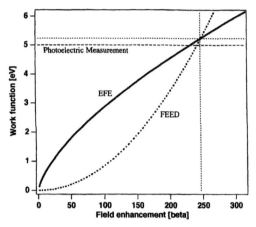

Fig. 1. Graph representing the relation between ϕ and β as obtained by measuring EFE and FEED from a fine grained CVD diamond film. The dashed line represents the data from the width of the FEED distribution and the solid line reflects the FN measurements. The intersection of the two curves gives the (ϕ, β) values for this particular sample (5.2 eV/247). It has to be stressed that the EFE and the FEED measurements are performed at the same time and from the same emitter.

computed. Hence, FEED and I/V measurements are taken from the same emitter at the same time. The intersection of the curves gives a pair ϕ and β which at the same time fulfill the FN equation and the equation for the width of the peak. Hence, at the intersection of the curves one can deduce the work function ϕ and the field enhancement factor β for this specific example. In addition, the global work function (measured over mm^2) can be measured by UPS and is marked in Fig. 1 as well. We have presented this powerful technique in different papers [17–19]. It allows us to determine a possible energy loss by measuring the energy of the electrons and from the shape of the distribution, it is possible to gain insight in the emission mechanism. So far, on all samples we have performed FEED measurements we found emission from the Fermi level, along with high local fields at the emission site. The responsible emission mechanism was found to be classical electron tunneling known from metals. Field enhancement at the emission site stems from clusters, tips or protrusions which are likely to occur during the deposition of electron emitting carbon films.

Even though we observed emission from the Fermi level, shifts in the FEED spectra could be observed in some cases. They also show up in the Fowler-Nordheim plot as a saturation at higher fields. The shift reflects a potential drop in the emission process and can in most cases be related to bulk resistivities of the emitting structure. However, there is an other mechanism which can lead to ohmic loss and hence to a shift of the FEED spectrum which to our knowledge was not discussed so far in the literature. It is known from investigations on diamond surfaces that a hydrogen saturation leads to some kind of surface conductivity [20–22]. It is partly due to this conductivity that electron microscopy on CVD-grown diamond is possible. Furthermore, this effect is used to produce one of the best performing FET's with diamond [21]. The conductive channel is a hydrogen saturated diamond surface with an aluminum electrode as the gate. The operation of these FET's shows that the surface conductivity is p-type [21]. Little is known about the mechanism of conduction except that hydrogen plays an important role.

In this paper we will discuss the deposition of nanotube films, analyze their emission properties and compare it to CVD grown diamond films. In a second part we will present a model experiment, showing that surface resistivity is an important issue when dealing with the electron emission from nanoclustered carbon films.

2. Experimental

FEED measurements were performed in an Omicron instrument equipped with an X-ray Photoelectron (XPS) and an Ultraviolet Photoelectron (UPS) Spectrometer. At a distance of 50 μm above the sample, a copper TEM mesh is placed and used as an extraction anode. The insulation between the grid and the anode is realized by Kapton® foil. By biasing the sample to a negative potential (Keithley 237) while keeping the grid at ground potential, electron field emission occurs and the energy of the electrons are measured with the hemispherical energy analyzer of the spectrometer [23]. All measurements were performed at a pressure below 10^{-9} mbar. The sample can be rotated in an azimuthal as well as in a polar direction for X-ray photoelectron diffraction [24]. This happened to be very important, as the electrons emitted from the surface may strike the bars of the copper mesh and produce a broad spectrum of secondary electrons. The contribution of these electrons to the FEED spectra can be minimized by carefully adjusting the sample position with respect to the entrance of the energy analyzer. FEED spectra were taken with a resolution of 70 meV, given by the energy analyzer and, as fast as possible, typically in few seconds. The reason is that the energy shift is dependent on the emission current, and hence a fluctuating current broadens the FEED spectra. Alternatively, I/V curves could be measured using a polished stainless steel sphere of 4 mm diameter mounted on a linear piezo drive.

CVD diamond and nanotube films were grown on silicon substrates via microwave plasma CVD in a tubular deposition system (2.45 GHz) at a gas pressure of 40 mbar and a substrate temperature of 840–1000°C from a CH_4/H_2 (2%/98%) gas mixture [18]. The growth parameters used are standard growth conditions for the deposition of CVD diamond films. For the growth of nanotube films,

metal was deposited either by sputter coating a very thin film of Ni (300–400 Å) or by spraying $Fe(NO_3)_3$ dissolved in ethanol onto the silicon substrate. After introducing the sample into the growth chamber $Fe(NO_3)_3$ is chemically reduced by the high concentration of atomic hydrogen to metallic Fe and forms little clusters due to the increased mobility at 900°C on the surface. A similar effect is observed on the sputter deposited Ni cluster. Any oxygen impurities are removed and the continuous thin layer of Ni forms little islands. These Ni or Fe clusters act as catalytic growth centres for nanotubes. A typical growth time was 15 min.

Electron emission from a metallic cluster on a diamond surface was achieved by fixing the cluster with a tweezers in the middle of the surface (3×4 mm^2) under the light microscope. The grain size was adjusted to be below 5 μm. The so prepared diamond was mounted on a special sample holder. Electrical contact was established by using silver paint to contact the metallic substrate holder and the diamond surface from the side. No direct contact was established to the grain itself except from the hydrogenated conductive surface layer. The diamond fixed on the sample holder could be transferred to an ECR plasma chamber to treat the surface by an oxygen microwave (2.45 GHz) plasma at 10^{-4} mbar with an RF bias of -50 V applied to the substrate for 90 s. Alternatively, the surface could be bombarded and graphitized by Ar$^+$-ions with an energy up to 2 kV. It has to be stressed that neither the grains nor the diamond itself had to be removed from the substrate holder for these treatments.

3. Results and discussion

3.1. Emission from nanotube films

Nanotube films were grown on silicon substrate by using a microwave or a hot filament reactor. The pretreatment has already been discussed in the previous paragraph. A main difference lies in the fact that atomic hydrogen is much more abundant in the microwave plasma than in a hot filament reactor. Hence, the growth conditions has to be tuned in a way that etching and growth of the film are in balance. Figure 2 shows a HRSEM picture of a nanotube film. The picture at the top shows phase pure nanotubes in a "spaghetti-like" structure, and at the bottom an image is presented where nanotubes grow from a metallic cluster. At higher resolution the diameter of the individual nanotubes can be determined to be in the range of 20–60 nm while its length can be as long as 100 μm. Nanotubes uniformly grow over the whole substrate and adhere quite well to it. These films are very well suited for electron field emission. In Fig. 3 we show an I/V characteristics with the inset being the Fowler-Nordheim plot. Emission starts at fields around 2.6 V/μm. The emission mechanism from such films is clearly governed by the field enhancement at the apex of the nanotubes. It is

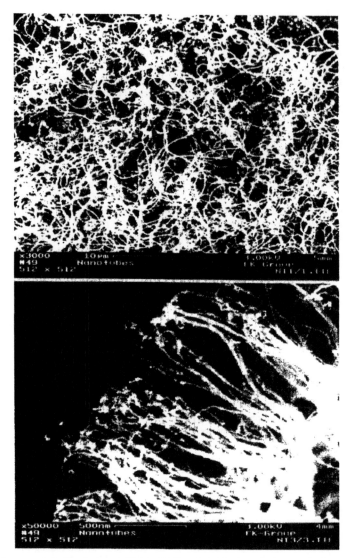

Fig. 2. HRSEM image from a nanotube film. At the top the spagetthi-like structure is shown and at the bottom the growth from one metallic cluster is shown.

interesting to note that a high density of nanotubes at the film surface does not give a high emission site density on a phosphor screen during field emission. This can easily be understood in terms of screening out the electrical field by too many nanotubes. Our process cannot align the nanotubes perpendicular to the surface in a controlled distance, which would certainly be the most ideal emitting structure. However, statistically a few nanotubes protrudes out of the surface and govern the emission mechanism. Fig. 4 shows the emission as monitored on a phosphor screen. The emission site density increases with the applied field and reaches approx. 10^4 at 3–4 V/μm when it becomes rather difficult to count the individual sites.

FEED measurements on these films revealed that the work function is in fact in the 5 eV range with a corresponding high field enhancement at the apex of the nanotubes. Figure 5 presents such a measurement. The electron emission originates at the Fermi level and is very well described and fitted by using classical field emission theory for metals.

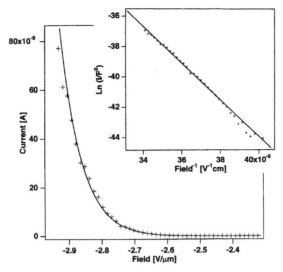

Fig. 3. I/V measurements on a nanotube film with the inset being the Fowler-Nordheim representation of the data. Emission starts at 2.6 V/μm.

It is interesting to compare field emission from nanotubes with the emission from CVD diamond films in view of the emission mechanism. Our measurements on a variety of CVD deposited films show that the emitting carbon structure is sp^2 bonded. The electrons are emitted from the Fermi level and the corresponding FEED spectra are very similar to what was discussed for the nanotube films. So far, we can explain all our results by assuming field enhancement by the geometrical effect. Even though, CVD diamond films do not show nanotubes at their surface, they posses very sharp, flake-like graphitic structures which easily account for the field enhancement. It is important to note that good emitting CVD diamond films show low quality diamond signal in the Raman spectrum.

3.2. Emission due to surface conductivity

Even though, there are different opinions about the emission mechanism from carbon films, there is agreement on the fact that good emitting material shows a

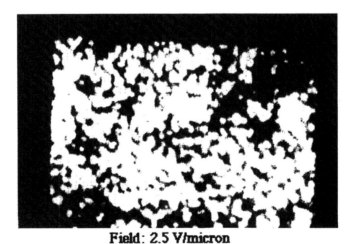

Field: 2.5 V/micron

Fig. 4. Electron field emission as monitored by a phosphorous screen at 3 V/μm. The size of the image is 1.53 cm².

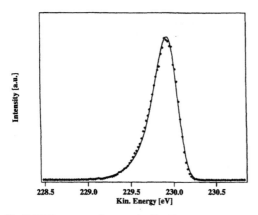

Fig. 5. FEED spectrum of a nanotube film. The energy is given in kinetic energy with the zero being the Fermi energy. The applied potential was 230 V and hence electrons are emitted from the Fermi level.

nanocrystalline structure. However, such a surface has a high surface-atom to bulk-atom ratio and it can be expected that surface treatment has a pronounced effect on the emission properties. In order to test this assumption we have investigated in a model experiment the emission from a micrometer sized cluster fixed on a previously hydrogen saturated diamond (100) surface. The electron supply to the cluster is maintained by surface conductivity only.

Figure 6 illustrates a FEED spectrum from hydrogenated surface at 800 V bias. The broad contribution at lower energies is due to electrons which on the way to the analyzer strike the copper grid and produce secondary electrons. This was confirmed by tilting the sample with

the phi-thetha manipulator while observing changes at lower energies. The cut off at higher energies is given by electrons which are emitted without losses from the indium grain. However, the cut off arises at very low energies. When applying a voltage difference of 800 V between anode and cathode, one would expect a cut-off at the same energy. However, such an energy could not be observed. Instead, we measured cut-off energies as low as 250 eV. Obviously, electrons do not see the full applied potential but just part of it. This is a direct proof that a hydrogenated diamond surface is conductive even though the resistivity is rather high, may be even not ohmic.

By an oxygen plasma treatment (ECR plasma) the diamond surface becomes insulating an the emission ceases. However, by a subsequent Ar^+ sputtering of the surface (1 kV) the insulating sp^3 configuration of the C-atom at the surface is changed into sp^2 and the surface becomes conductive and electron emission is observed again. However, the surface resistivity depends on the sputtering time and hence the shift in the FEED spectra with respect to the Fermi level decreases with sputtering time. In Fig. 7 we present a FEED spectra of the sputtered surface after 15 min. The emission originates near the Fermi level. However, a shift in the peak can still be observed, which depends on the applied potential. Figure 8 shows the dependence on the applied potential and the emitted current. This figure clearly shows that the shift follows a linear scaling law with the emitted current but not with the applied potential. Hence the surface resistivity is indeed ohmic and can be determined to be 22 MΩ after 15 min of sputtering. The dependence on the applied voltage can easily be computed when assuming a resistivity in series with the emission process and using Ohm's law.

This model experiment has important consequences on

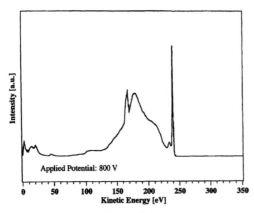

Fig. 6. FEED spectrum for an applied potential of 800 V. The broad features at lower energies are secondary electrons produced by the extraction grid 50 μm in front of the surface. The maximum kinetic energy of the field emitted electrons (250 V) is much lower than the applied bias (800 V).

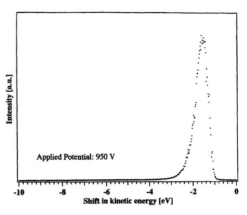

Fig. 7. Emission from the sputtered diamond surface at a sample bias of 950 V. Electrons are emitted with an energy close to the Fermi level. The shift is 1.7 eV.

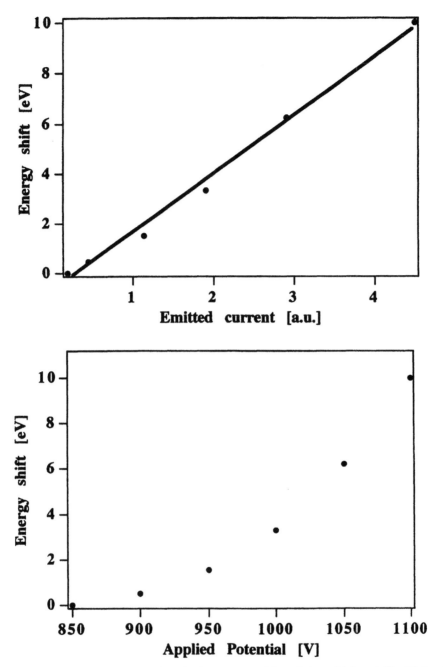

Fig. 8. Observed energy shift in the FEED spectra vs the emitted current (top) and the applied potential (bottom). The shift depends linearly on the emitted current and hence the surface resistivity is ohmic. The dependence on the applied potential is given by Ohm's law assuming electron emission with a series resistivity being the surface.

the electron emission from nanostructured carbon films. First of all, we have shown that emission is made possible by surface conductivity, either by hydrogen saturation of sp^3 or by sp^2 at the surface. This explains why surface treatment after the deposition of carbon films can alter the electron emission. While a hydrogen plasma treatment can be beneficial to connect through surface conductivity sp^2 region which otherwise would be lost for electron emission, an annealing step can be beneficial as well, when building up sp^2 conductive layers at the surface. Hence, surface treatment of carbon films is certainly an important issue. The success of a treatment however, depends on the nanostructure of the sample under consideration. It can be beneficial but also kill the emission. This explains the controversial results encountered in literature about the hydrogen plasma treatment of carbon films. Secondly, our model experiment has shown that surface resistivity leads to a shift in the FEED spectra due to an ohmic loss across the surface. While hydrogen-saturated surfaces (diamond or tetrahedral carbon) show a rather high resistivity, sp^2 layers on an otherwise insulating surface leads to a conductivity which depends on the thickness of the layer. Nanocarbon films have a high effective surface and hence shifts in the FEED spectra can be expected to occur. These shifts are due to an ohmic loss, as discussed, and not due to injection of electrons in the conduction band of the nanocrystals. Hence, a shift in the FEED spectra can not be interpreted as a proof for injection. The problem whether injection or an ohmic loss causes the shift, needs a more detailed analysis of the shape of the FEED spectra.

4. Conclusions

We have presented a deposition method to produce flat nanotube films on silicon surfaces. These films are very phase pure (just nanotubes, no graphitic clusters) except from the metallic catalyst and are very well suited for field emission applications. We observed electron emission with an onset of 2.6 V/μm (for 1 nA emission current) and an emission site density reaching $10^4/cm^2$ at 3–4 V/μm. Field emission spectroscopy revealed that the electrons are emitted from the Fermi level. The electrical field at the emission site is amplified by the tip geometry of the nanotube and the work function is in the 5 eV range. The observed results are very similar to what we measured on CVD diamond films showing a ballas-like morphology.

By fixing a micrometer sized metallic cluster on a differently prepared diamond (100) surface (hydrogen saturated, oxygen saturated, sputtered) we could show that in the FEED spectra large energy shifts could be observed due to the resistivity of the surface where the applied potential is partly lost. This experiment has important consequences for the understanding of electron emission from nanocrystalline CVD carbon films.

Acknowledgements

Part of this work was financed by the Swiss Priority Program on Materials (PPM) and by the Swiss National Science Foundation (NFP36).

References

[1] Amaratunga GAJ, Silva SRP. Appl Phys Lett 1996;68:2529.
[2] Satyanarayana BA, Hart A, Milne WI, Robertson J. Appl Phys Lett 1997;71:1430.
[3] Gröning O, Küttel OM, Gröning P, Schlapbach L. Appl Surf Sci 1997;111:135.
[4] Okano K, Koizumi S, Silva SRP, Amaratunga GAJ. Nature 1996;381:140.
[5] Geis MW, Twichell JC, Macaulay J, Okano K. Appl Phys Lett 1995;67:1328.
[6] Geis MW, Twichell JC, Lyszcarz TM. J Vac Sci Technol 1996;B14:2060.
[7] Huang ZH, Cutler PH, Miskovsky NM, Sullivan TE. Appl Phys Lett 1994;65:2562.
[8] An overview is given in: Latham R, editor. High voltage vacuum insulation, Academic Press, 1995 and references therein. Xu NS, Latham RV, Tzeng Y. Electron Lett 1993;29, 1596.
[9] Schlesser R, McClure MT, Choi WB, Hren JJ, Sitar Z. Appl Phys Lett 1997;70:1596.
[10] Robertson J. Diamond Related Mat 1996;5:797.
[11] Robertson J. Phys Rev 1996;B53:16302.
[12] Diamond WT. J Vac Sci Technol 1998;A16:707.
[13] Diamond WT. J Vac Sci Technol 1998;A16:720.
[14] Henderson JE, Badgley RE. Phys Rev 1931;38:540.
[15] Gadzuk JW, Plummer EW. Rev Mod Phys 1973;45:487.
[16] Bandis C, Pate BB. Appl Phys Lett 1996;69:366.
[17] Gröning O, Küttel OM, Nilsson L, Diederich L, Schlapbach L. Appl Phys Lett 1997;71:2253.
[18] Küttel OM, Gröning O, Nilsson L, Diederich L, Schlapbach L. 1st Specialist Meeting on Amorphous Carbon, Cambridge, UK, 1997, in press.
[19] Küttel OM, Gröning O, Nilsson L, Diederich L, Schlapbach, L. MRS Fall Meeting, Boston, 1997.
[20] Heiland G. Zeitschrift für Physik 1957;148:15.
[21] Gluche P, Aleskov A, Vescan A, Bert W, Kohn E. IEEE Electron Device Lett, November 1997.
[22] Kawarada H, Aoki M, Ito M. Appl Phys Lett 1994;65:1563.
[23] Gröning O, Küttel OM, Gröning P, Schlapbach L. Appl Phys Lett, in press.
[24] Küttel OM, Agostino RG, Fasel R, Osterwalder J, Schlapbach L. Surf Sci 1994;312:131.

Pergamon

Carbon 37 (1999) 753–757

CARBON

Field emission of nitrogenated amorphous carbon films

U. Hoffmann*, A. Weber, C.-P. Klages, T. Matthée

Fraunhofer-Institute for Surface and Thin Film Technology, Bienroder Weg 54e, 38108 Braunschweig, Germany

Received 16 June 1998; accepted 3 October 1998

Abstract

Field emitter films based on nitrogenated amorphous carbon (a-C:N) were deposited on different chromium patterns on glass by sputtering of graphite employing an electron cyclotron resonance plasma as argon and nitrogen ion source. The a-C:N films contain between 0.6 and 21 at.% nitrogen. All films have a low resistivity (<0.1 Ωcm) and a microhardness of about 15 GPa indicating a high content of sp^2 bonds. The vacuum electronic properties of the films were checked in an UHV chamber in a plane to plane set-up. To localize the emission sites the excitation of a low voltage phosphor (ZnO:Zn) was monitored by a CCD camera. After an activation by vacuum arc discharges emission of electrons occurred at macroscopic electrical fields as low as 3.2 V/μm. The discharges generate delaminated a-C:N film fragments that bear the FE current due to field enhancement. Discharge and therefore FE only took place at the edges of the emitter stripes due to a macroscopic field enhancement. Two kinds of activation were found leading to different microstructures of the emitter and different FE characteristic. The influence of substrate bias, nitrogen content, film thickness and emitter geometry on the field emission was also surveyed. © 1999 Elsevier Science Ltd. All rights reserved.

Keywords: A. Carbon films; B. Plasma sputtering; D. Electronic properties; Field emission

1. Introduction

Electron field emission (FE) from carbon based materials has become an area of great interest for application, e.g. in field emission displays (FEDs). Field emission at macroscopic fields of about 10 V/μm was found with different carbon based films like diamond [1], nanocrystalline diamond containing metal film deposited by electroplating [2] or dielectrophoresis [3], amorphous carbon (a-C and ta-C) [4], nitrogenated a-C and ta-C [5,6], hydrogenated amorphous carbon (a-C:H) [7] and carbon nanotubes [8]. The mechanism of cold electron emission from these materials is still not clear. Stable FE was found in most cases [9,10,5] only after activating the films by vacuum arc discharge or another forming process, e.g. repeated voltage ramps [6]. Incorporation of small amounts of nitrogen in a-C and ta-C films seems to improve the FE-characteristic [5,6] possibly due to a doping of the films [11]. Although FEDs are usually named as the application for field emission from carbon based materials, the commonly used substrate is unstructured silicon. The

I–V characteristics are often measured in a point to plane configuration with a needle probe [10,12] or the tip of a special STM [9] as anode. This setup provides a good spatial resolution for the FE characteristic of a film, but it is not trivial to transfer the results to a plane to plane setup as in a FED [13].

This paper compares the FE from different a-C:N films. Two kinds of activation will be presented and their influence as well as the influence of the substrate bias, the nitrogen content and the film thickness on the FE characteristic will be discussed.

2. Experimental

Emitter structures were prepared by conventional photolithography methods with only one patterning step. First a 150 nm thick chromium layer was deposited by e-beam evaporation onto a photoresist pattern on a glass substrate. Then the preparation of thin a-C:N films was performed by sputtering graphite targets using an electron cyclotron resonance plasma. An ECR microwave (MW) plasma generated in an ASTeX HPM/M source was extracted from the resonance cylinder to a graphite target. A flow of 20 sccm argon–nitrogen mixture at a pressure of 0.1 Pa

*Corresponding author. Tel.: +49 531 2155670; fax: +49 531 2155900.
E-mail address: hof@ist.fhg.de (U. Hoffmann)

0008-6223/99/$ – see front matter © 1999 Elsevier Science Ltd. All rights reserved.
PII: S0008-6223(98)00266-8

was used as sputter gas. The deposition set-up was described elsewhere [14]. Nitrogen content in the gas was varied between 0.001% and 25% N_2. The MW power was set to 300 W and the target bias was −1200 V. Substrate bias was varied from −2 V to −100 V. The deposition rate was 5 nm/min and the thickness of the surveyed layers was varied between 30 and 300 nm. Figure 1 shows some of the investigated structures.

The field emission properties were investigated in a plane to plane set-up in a vacuum chamber with base pressure smaller than 10^{-5} Pa. Current voltage characteristics were measured using a computer controlled voltage source (Stanford Research PS325) and an ammeter (Keithley 2000 Multimeter) [5]. Two types of anodes were used: glass with sputtered indium tin oxide (ITO) and a fluorescent screen with a zinc oxide: zinc (ZnO:Zn) phosphor on ITO on glass. The latter was used to display the spatial and temporal distribution of the emission by the excitation of the ZnO:Zn. The spacers were Kapton pads (60 μm high) which had no contact with the ITO or chromium layers. The set-up for the current–voltage (I–V) measurements was tested once with a chromium on glass sample and an ITO Anode. Up to 1500 V no currents and discharges were found. Each a-C:N coated sample showing voltage-dependent current was tested by applying inverse voltage to the electrodes (negative voltage to the ITO). By using a mobile microscope equipped with a CCD camera it was not only possible to see sparks at vacuum arc discharge and the emission sites during stable emission, but also to record the images by a S-VHS video recorder.

Film composition was analysed by secondary ion mass spectrometry (SIMS) and electron probe microanalysis (EPMA), and the crystallinity by X-ray diffraction (XRD) and high resolution transmission electron microscopy (HRTEM). Microstructures of the emitter sites before and after FE measurements were investigated with a SEM (Jeol, Type 6300F) provided with a field emission cathode for a good resolution even for low electron energies. Ionization energy of the a-C:N films was determined by

UV photo electron spectroscopy (UPS). Surface morphology and roughness was characterized with a scanning force microscope (AFM) and the microhardness was measured with a nanometer indentation device. Film resistivity was checked by a four-point probe ohmmeter.

3. Results and discussion

The influence of nitrogen on microhardness, RMS roughness and resistivity of the a-C:N films was investigated. Nitrogen content was measured with SIMS for small and EPMA for larger N-contents. All films contain less than 0.3 at.% oxygen and between 1 and 4 at.% hydrogen. The RMS-roughness measured with an AFM decreases from about 36 Å for small N-contents to very smooth films with roughness of 2 Å for N-contents of more than 13 at.%. The microhardness investigated with a nanoindentor on thick (≥400 nm) films shows no clear dependence on the N-content. A microhardness of about (15±5) GPa indicates a sp^2 content of more than 80% [15]. This might also be the reason for the low resistivity of the films. There seems to be a minimum in resistivity (0.015 Ωcm) for a nitrogen content of about 5 at.%, but the variation is inconsiderable. All films have a low resistivity (<0.1 Ωcm), so the electron transport from the back contact to the emission site is not the limiting factor for FE as it is in insulating films. Studies with XRD and HRTEM show no graphitic clusters or other kind of ordered structures, so the CN-films are amorphous.

Figure 2 shows an I(V) plot of the first measurement of an a-C:N film. The pressure at the beginning of the test was smaller than 10^{-5} Pa. The emission current starts to rise at 440 V corresponding to a macroscopic field of about 7 V/μm. The current does not only rise considerably while further increasing the voltage but even at constant voltage

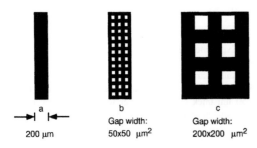

Fig. 1. Top-view of 3 of 10 tested emitter structures: (a) a strip with 200 μm width; (b)+(c) grids with 50×50 μm² and 200×200 μm² gaps. The black areas are a-C:N on chromium on glass, the white ones are only glass.

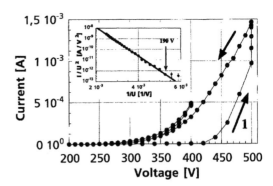

Fig. 2. I(V) plot of an a-C:N film showing *slow* activation starting at 440 V (1). Spacing between emitter stripes and fluorescent screen was 60 μm. The inset shows a FN-like plot indicating FE down to 190 V.

of 500 V. After the activation FE is stable down to 190 V corresponding to a macroscopic field of about 3.2 V/μm. This kind of activation shall be called a *slow* activation because it is correlated to a lot of microdischarges occurring in succession during some ten seconds or even a few minutes. The first discharges raise the local pressure and might lead to some particles in their surrounding. This leads to a higher discharge probability and the next discharges took place near to the first. This slow activation took typically place at voltages of about 500 V. The second kind of activation shall be called *fast* because it is correlated to a lot of small discharges in less than 40 milliseconds (only one or two video frames). Fast activation took place at high voltages of typically more than 1300 V and lead to an automatic shut off of the power supply. The microdischarges create craters wich are typical 10 μm in diameter and appear only at the edges of the emitter rows due to a macroscopic field enhancement which will be discussed later. The craters are separated by undamaged areas typical 10 μm long as well. Figure 3 shows the microstructure of craters after a slow (a) and a fast (b) activation. In both cases the chromium film was melted and evaporated by the discharge and even the glass substrate was damaged as detected with an AFM. At slow (low voltage) activation (Fig. 3(a)) the a-C:N film delaminates continuously from the chromium around the crater, often containing fragments out of the crater area. At fast (high voltage) activation (Fig. 3(b)) most of the a-C:N film of the crater and also of the surrounding area was removed so that free chromium surface appeares. But also for fast activation, delaminated film fragments can often be found near the crater. In the author's opinion, the field emitted electrons originate from these fragments because the edges of the delaminated a-C:N films might have high local microscopic field enhancement factors (β_μ). The real microstructure and therefore the electric field at the emitter

can be assessed only roughly. The local electric field (E_l) at the emitter should be written [16]

$$E_l = \beta_\mu \beta_g V/d$$

where V is the applied voltage, d is the gap between anode and cathode, β_μ and β_g are the field enhancement related to microscopic geometry of the emitter and the gross geometry of the gap, respectively. A value for β_μ can be roughly estimated as: $\beta_\mu \approx h/r+2$ [17] for a cylinder (height h) capped with a hemisphere (radius r). For our emitter we assume from SEM images $h=5$ μm and r is equal to or smaller than half of the a-CN film thickness. So for a typical thickness of 50 nm we have: $\beta_\mu \approx 200$. A macroscopic field enhancement (β_g) took place at the edges of the emitter lines and therefore the discharges always occur there forming the emission sites. A value for β_g is difficult to estimate. Tests with different types of grid-structures as emitter line show that for discharges at inner edges the spacing between two lines with the same potential has to be larger than the gap between emitter and anode (60 μm). For a grid with 50 μm spacing discharge followed by FE was only found at the outer edges, even with 100 μm spacing FE rarely occured at the inner edges. Apart from the field enhancement some other results point out that the delaminated a-CN films bear the FE current: in our plane to plane set-up not all craters show FE. This can be explained by the different and random β_μ values of the delaminated films but not only due to the forming of another emitting material, e.g. chromium carbide. Of course it is possible that the structure and composition of the delaminated a-C:N film changes, e.g. the sp^2-fraction (due to graphitization) or the nitrogen content, but the field enhancement is the dominating factor and we still find a influence of the nitrogen content on the FE characteristic after activation. After wiping away the delaminated films

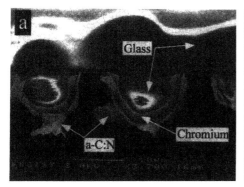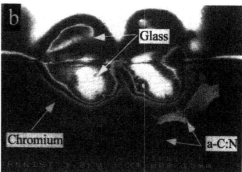

Fig. 3. SEM images of the microstructure of craters after a slow (a) and a fast (b) activation. It is a top-view on the emitter structure like in Fig. 1. The SEM images are taken with low electron energy (3 kV) because at higher energies the thin (50 nm) delaminated a-C:N films, which bear the FE current, are nearly invisible due to irradiation.

by rubbing one emitter structure with a piece of acetone soaked paper, this structure has to be activated by discharges for FE again. An adjacent unchanged emitter structure at the same sample shows a constant FE characteristic. All this indicates that parts of the delaminated films bear the FE current due to the enhanced electric field.

Also the microdischarges produce emitters with a random geometry a careful comparison of the FE characteristic of the different a-C:N film makes sense because the great number (about 100) of simultaneously measured emission sites on one emitter structure averages the different β_μ values. In our plane to plane set-up only comparably strong emitter contribute to FE current. To compare the different a-C:N films we use the voltage ($V_{10\mu A}$) for a total FE current of 10 μA for the same emitter structures. From varying the substrate bias during a-C:N deposition we found an improved FE for small bias values (= −20 V). The nitrogen content has only a small influence on $V_{10\mu A}$ (Fig. 4). Varying the N-content in the films from 0.6 at % to 21 at.% we found a minimum for $V_{10\mu A}$ at 8 at % N, but the influence is weak. Thin films have smaller $V_{10\mu A}$ not due to the smaller resistance (all films have a low resistivity) but due to the higher β_μ values ($\beta_\mu \approx h/r+2$ and $r \leq$ half film thickness). The kind of activation influences the FE characteristic, too. On nine samples we have emitter structures with slow as well as ones with fast activation. For one sample $V_{10\mu A}$ was a little higher for a slow activated structure, for all other samples the slow activated structures have partly notable lower $V_{10\mu A}$ values. The reason might be the different microstructure of the discharge craters (see Fig. 3). For slow activation more craters are surrounded by delaminated films, so high β_μ values are more frequent.

After the first activation the emission characteristic for an emitter structure is constant even after storing the sample in air for more than 1 month. Figure 5 shows a photo of the excited phosphor screen and a schematic drawing of the emitter. For this structure discharges and FE were found at the outer and the inner edges because β_g is nearly the same for both.

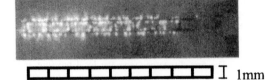

Fig. 5. Excitation of the phosphor screen by electron emitted from a-C:N and a schematic drawing of the emitter structure. Applied voltage was 400 V, current 800 μA and cathode-screen spacing 60 μm.

The duration of the FE measurement has no obvious influence on the microstructure of the emitter. The craters with delaminated a-C:N films look similar for standard measurement as shown in Fig. 2 (duration about 10 min.) and for long time measurements (about 90 min.) with currents of 400 μA at 400 V for a 36 mm long grid (see Fig. 1(b)).

4. Conclusion

Field emission from different a-C:N films was investigated with structured emitter stripes and a plane phosphor screen as anode, similar to the set-up in a carbon based FED. Activation by vacuum arc discharges at the edges of the emitter stripes leads to the forming of craters surrounded by delaminated a-C:N film fragments. These fragments bear the FE current due to the field enhancement caused by their microstructure and the field enhancement due to gross geometry of the gap. In addition UPS measurements indicate no negative electron affinity for the films and a ionization energy of 4.6 eV. After the activation the FE is stable and the macroscopic onset field is only 3.2 V/μm. Best FE was found for slow activated thin films, deposited with −20 V substrate bias, containing about 8 at.% nitrogen.

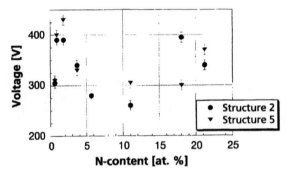

Fig. 4. Influence of the N-content on $V_{10\mu A}$

References

[1] Wang C, Garcia A, Ingram DC, Lake M, Kordesch ME. Electron Lett 1991;27:477.

[2] Weber A, Hoffmann U, Löhken, T, et al. SID International Symposium, Digest of technical papers, 1997:591.

[3] Choi WB, Cuomo JJ, Zhirnov VV, Myers AF, Hren JJ. Appl Phys Lett 1996;68:720.

[4] Kumar N, Schmidt H, Xie C. Solid State Technology 1995;5:71.

[5] Hoffmann U, Weber A, Löhken T, Klages CP, Spaeth C, Richter F. Diam Rel Mat 1998;7:682.

[6] Satyanarayana BS, Hart A, Milne WI, Robertson J. Diam Rel Mat 1998;7:656.

[7] Amaratunga GAJ, Silva SRP. Appl Phys Lett 1996;68:2529.

[8] Gulyaev YuV, Chernozatonskii LA, Kosakovskaja ZJa, Sinitsyn NI, Torgashov GV, Zakharchenko YuF. J Vac Sci Technol 1995;B 13:435.

[9] Gröning O, Küttel OM, Schaller E, Gröning P, Schlapbach L. Appl Phys Lett 1996;69:476.

[10] Talin AA, Felter TE, Friedmann TA, Sullivan JP, Siegel MP. J Vac Sci Technol 1996;A 14:1719.

[11] Robertson J, Rutter MJ. Diam Rel Mat 1998;7:620.

[12] Göhl A, Habermann T, Müller G, et al. Diam Rel Mat 1998;7:666.

[13] Diamond WT. J Vac Sci Technol 1998;A 16:707.

[14] Weber A, Hoffmann U, Klages CP. J Vac Sci Technol A 1998;16:919.

[15] Weiler M, Sattel S, Giessen T, Jung K, Ehrhardt H, Veerasamy VS, Robertson J. Phys Rev B 1996;53:1594.

[16] Farrall GA. In: Lafferty JM (editor). Vacuum arcs, theory and application. New York: Wiley, 1980.

[17] Chatterton PA. Proc Phys Soc 1966;88:231.

Pergamon

Carbon 37 (1999) 759–763

CARBON

Electron field emission from diamond and diamond-like carbon for field emission displays

J. Robertson*

Engineering Department, Cambridge University, Cambridge CB2 1PZ, UK

Received 16 June 1998; accepted 3 October 1998

Abstract

It is shown that the facile electron field emission from diamond and diamond-like carbon occurs because surface groups such as C-H can produce large changes in electron affinity, so that electric fields from the anode can be focused towards unhydrogenated surface areas of high affinity, the fields ending on negative charges in an underlying depletion layer. The resulting downwards band bending creates very large fields which cause Fowler-Nordheim emission, while not exceeding the material's breakdown field, which is the highest for any solid. © 1999 Elsevier Science Ltd. All rights reserved.

Keywords: A. Diamond; Diamond-like carbon; D. Field emission; Surface properties

1. Introduction

Flat panel displays with an image quality similar to cathode ray tubes can, in principle, be produced by field emission displays (FEDs). Existing FEDs use the emission from sharp 'Spindt' tips of materials such as Mo or Si. Future FEDs may use flat film cathodes of easy electron emitters such as diamond, diamond-like carbon (DLC) or nano-structured carbons. DLC has the advantage that it can be deposited at room temperature over large areas onto glass substrates. The original work on emission from diamond was motivated by the discovery of its negative electron affinity (NEA) [1,2]. The ease of emission from the large variety of carbon-based materials suggests that the NEA is not a prerequisite, and more general emission models are desired for these systems.

There have been numerous studies of field emission from diamond [3–7], DLC [8–10] and nanostructured carbon [10–14]. A display needs ~0.1 mA/cm² current density assuming an anode voltage of ~2 kV. A threshold field for 1 μA/cm² has been used as a merit parameter to distinguish various emitter materials. With a wide band gap of 5.5 eV, field emission from high quality diamond is actually quite difficult because of its high resistivity. Emission from polycrystalline CVD diamond is easier because of its conducting grain boundaries and it is found to vary inversely with the grain size [3]. The lowest

threshold fields were found by Talin et al. [6] for nanocrystalline diamond, and by Zhu et al. [7] for sintered diamond grits. Boron-doped diamond has a high threshold field for emission and this is lowered by (nominal) phosphorus doping and strikingly by nitrogen doping [4] to field as low as 0.5 V/μm.

DLC is a semiconductor whose band gap varies from 1 eV to 4 eV according to its sp³ content. Field emission from DLC is generally easier than for diamond. Tested in parallel plate geometries, the typical threshold emission field for DLC is 20–40 V/μm [8], which can be lowered to about 10 V/μm for more optimised films, and to 5 V/μm for nitrogen doped films [9]. For tetrahedral amorphous carbon or 'ta-C' deposited by filtered cathodic vacuum arc, the threshold field decreases from 25 to about 10 V/μm for the highest sp³ content [9]. Some other types of carbon can make good emitters. As points enhance the local field, emission is quite easy from carbon nanotubes. DeHeer et al. [11,12] found emission at 15 V/μm while Wang et al. [13] obtained emission at 0.8 V/μm.

For displays, it is now recognised that the threshold field parameter has limited value. Because of the exponential character of the Fowler-Nordheim equation, emission often occurs from particular sites on the film. A critical parameter is the emission site density, which should approach 10⁶ cm⁻² for a display quality material. The development by Coll et al. [10] of a nanostructured carbon with a 'coral-like' surface is important. This carbon has a surface with numerous coral-like protrusions with radius 3–7 nm, and

---

E-mail address: jr@eng.cam.ac.uk (J. Robertson)

gives emission at 5–10 V/μm with a very high site density.

In view of the NEA of diamond, it was first assumed that the easy field emission from carbon arose from a low electron affinity. However, closer inspection of the data shows that this is not correct. Field emission is a tunnelling process and the current density J (Am^{-2}) obeys the Fowler-Nordheim equation

$$J = aE^2 \cdot \exp\left(-\frac{b\phi^{3/2}}{\beta E} \right) \tag{1}$$

where ϕ (eV) is the barrier height, E is the applied field (V/m), β is a dimensionless geometric field enhancement factor and a and b are constants, $b = 6.8.10^9$. Plotting the field emission current densities for diamond according to the Fowler-Nordheim equation gives an effective barrier ϕ of 0.1 to 0.4 eV for no field enhancement ($\beta = 1$). A similar plot for ta-C gives barrier heights for $\beta = 1$ of 0.04 to 0.1 eV and down to 0.01 eV for N-doped films (Fig. 1) [9]. These values are impossibly small.

The electron energy distribution (EED) of the emitted electrons is a key means to test various emission mecha-

nisms, because it distinguishes the states the electrons come from, whether there is a voltage drop across the film by referencing to the Fermi level of the back contact (Fig. 2a) or if there is hot electron emission which reveals itself as a broadening on the high energy side (Fig. 2b). In addition, the width of the EED is proportional to the local emission field. Bandis and Pate [15] made simultaneous field and photoemission measurements and observed that the field emitted electrons from a single crystal type 2B diamond came from the valence band, not from the conduction band. Also the EED width gave a very high local field of order 2000 V/μm for applied fields of 50 V/μm. Groning et al. [16] observed similar results for nitrogen alloyed a-C.

In summary, there are two possible models of field emission from carbon films; either the barrier to emission is very low or there are extremely high electric fields involved. In view of the low electron affinity (EA) of these systems, the low barrier model is naturally popular. However, the barrier values observed, ~0.1 eV, are unrealistic low. The alternative is that the barrier is ~5 eV and the fields are high, as suggested by the measured EEDs [15,16]. However, we must then explain the origin of high fields in both diamond and DLC; as DLC is smooth this precludes a general field enhancement at sharp points.

2. Electronic structure

A mechanism which resolves these problems and shows the uniqueness of carbon systems has recently been developed [17,18]. First we summarise the surface electronic structure of diamond and DLC. Fig. 3 shows the band diagram of diamond on a metal substrate. The surface is shown with a small or negative EA but there is a large barrier for electron injection into the conduction band from the back contact.

Figure 4 summarises the band line-ups for DLC [17,18]. The affinities of typical a-C:H with gaps of order 2.0 eV is 2.5–3.0 eV [17–19]. At the back contact, a-C:H on c-Si has little offset between conduction bands [20], Fig. 4. This band line-up means that the main barrier for electrons is at the front surface, at least 3.5 eV for the work function.

The bare surface of diamond has an EA of about 0.35 eV while the hydrogenated surface has a NEA of −1.3 eV [21,22] (Fig. 5). Thus, the valence band edge lies between about 6.0 eV and 4.0 eV below the vacuum level. Oxygen termination creates a large positive EA of ~2.6 eV. The chemical termination of the diamond surface has an effect on its EA, because the terminating groups, C-H, C-O etc. introduce a surface dipole layer which introduces a voltage step. For C-H surface, the H atom is slightly positive and the dipole reduces the affinity and makes it negative. For C-O or C=O groups, the oxygen is negative and the dipole makes the affinity more positive.

Field emission from semiconductors in general can

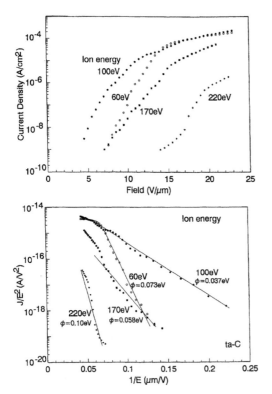

Fig. 1. Fowler-Nordheim plot of field emission currents from ta-C, showing effective barriers heights of 0.04 to 0.1 eV.

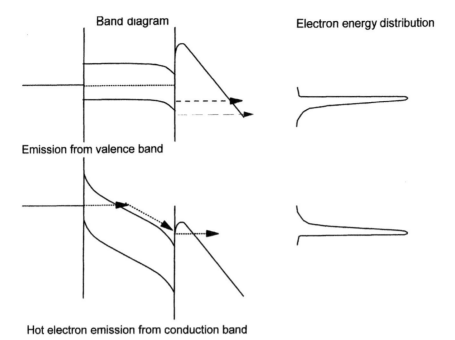

Fig. 2. (a) Electron energy distributions for field emission mainly from the valence band, showing a sharp high energy side and a width proportional to the local field. (b) Emission due to hot electrons, which shows a high energy tail.

originate from the valence band, gap states or the conduction band [23]. Previous models have attributed emission from diamond and DLC to the low affinity of diamond [2], the so-called antenna effect of conducting channels [24], emission from defect gap states [25], and band bending at depletion layers [5]. The defect density in DLC or within a diamond grain is very high so an external field ionises these defects in the surface layers and there is little field penetration into the film.

Emission from nitrogen or phosphorus doped diamond is a special case. The main barrier for electron emission for n-type diamond is the back contact (Fig. 3). It was first thought that nitrogen aids emission because it forms deep donor levels 1.4 eV below E_c and creates an ionised depletion layer which causes a strong upwards band bending towards the back contact. At sufficiently high donor concentrations, this band bending narrows the tunnel

Fig. 3. Schematic band model for diamond.

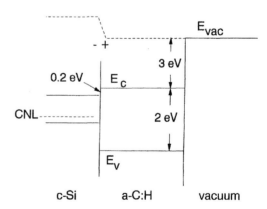

Fig. 4. Schematic band diagram of a-C:H on c-Si, from photo-emission data [16,18].

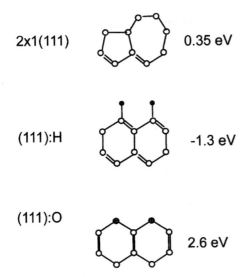

2x1(111)   0.35 eV

(111):H   -1.3 eV

(111):O   2.6 eV

Fig. 5. Electron affinities for diamond surfaces for various terminations [19,20].

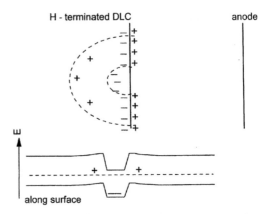

Fig. 6. Charge distribution and band diagrams at a DLC surface with small region of surface unterminated by hydrogen, with no applied field. The dipoles due to surface C-H groups are shown as + and −.

distance, aiding emission [5]. However, recent EEDs show that emission from Okano's samples occurs from 1.4 eV above E_v not 4.1 eV above E_v [26]. This suggests that the N has aggregated into A centres which have donor states 1.4 eV above E_v [27] (Fig. 3). These are less easily ionised in depletion layers and would require a Schottky barrier below 1.4 eV with an H-terminated diamond interface in the barrier.

Emission from DLC and polycrystalline diamond requires a different model. In DLC, the emission barrier of 3–4 eV is at the front surface (Fig. 4), while the Fowler-Nordheim slopes are very low, 0.01–0.1 eV. Polycrystalline diamond has a similar problem. The Fermi level of diamond tends to lie in the lower part of the gap, because grain boundary and vacancy related levels lie about 1.4 eV above the valence band edge (Fig. 3). Therefore, the barrier for emission from a diamond grain is still at the front surface and equals the work function, which is about 3–4.5 eV, depending on the surface termination. The field emission data in Fig. 1 give an effective barrier of 0.1 to 0.4 eV. Poly and nano-crystalline has a rough surface and heterogeneous bonding so they could have a large field enhancement factor. However, DLC has an extremely low surface roughness, often below 1 nm, so that there can be no geometric field enhancement (unless emission is all attributed to crater damage).

We propose the following mechanism for field emission. Consider a surface of diamond or DLC terminated by hydrogen, except for a small 10 nm region. Its EA is −1 eV except for the unterminated region where it is 1 eV. The

affinity changes are shown in Fig. 6 as arising from surface dipoles. The band bending at the unterminated region is compensated by charged defects in the underlying film. On the application of an anode field, instead of the field lines terminating in a depletion layer spread uniformly over the surface, they focus to the negative depletion layer under the unterminated region. This focusing greatly increases the field through the tunnel barrier (see Fig. 7). The field size is given by the potential step ΔV caused by differences

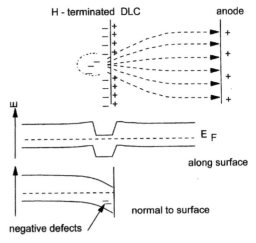

Fig. 7. With applied field. The field lines now end at negatively charge defects in a hemi-spherical depletion region behind the surface, causing strong downwards band bending.

Table 1
Comparison of breakdown fields

	Breakdown field, $V/\mu m$
Si	30
GaAs	40
SiC	300
diamond	1000
ta-C	~800

in termination, and the defect density, N, in the depletion layer,

$$E = \left(\frac{2N.e.\Delta V}{\varepsilon} \right)^{1/2} \qquad (2)$$

For $\Delta V = 2$ V, $N = 10^{25}$ m^{-3}, $\varepsilon_0 = 8.9.10^{-12}$ F/m, and $\varepsilon_r = 5$, then $E = 4.10^8$ V/m in the tunnel barrier. This size field is roughly that needed to account for the observed EEDs [16].

Emission occurs mainly from nanometer sized surface regions unterminated by H with more positive affinity, which causes a very strong downwards band-bending. In diamond, the emission will tend to come from the surfaces of more sp^2 bonded grain boundaries, as these are less likely to be H-terminated. In DLC, the model introduces large, inhomogeneous fields without roughness.

The mechanism is restricted to carbon systems because

1. changes in termination can create such large potential fluctuations, because of the high surface atom density,
2. the defect density in DLC and even in a diamond grain is comparatively large, so the depletion width is short and the fields are high,
3. exceptionally large fields can occur in carbon systems because of their high breakdown fields, of order 1000 $V/\mu m$ (Table 1) [28].

The mechanism is related to the antenna model of Latham [27], except in using a chemical rather than electrical inhomogeneity.

References

[1] Himpsel FJ, Knapp JS, VanVechten JA, Eastman DE. Phys Rev B 1979;20:624.
[2] Geis MW, et al. IEEE Trans ED Lett 1991;12:456.
[3] Zhu W, Kochanski GP, Jin S, Seibles L. J Appl Phys 1995;78:2707.
[4] Okano K, Koizumi S, Silva SRP, Amaratunga GAJ. Nature 1996;381:140.
[5] Geis MW, Twichell JC, Lyszczarz TM. J Vac Sci Technol B 1996;14:2060.
[6] Talin AA, Pan LS, McCarty KF, Doerr HJ, Bunshah RF. Appl Phys Lett 1996;69:3842.
[7] Zhu W. Presented at Diamond Films '97. Edinburgh, UK.
[8] Amaratunga GAJ, Silva SRP. App Phys Lett 1996;68:2529.
[9] Satyanarayana BS, Hart A, Milne WI, Robertson J. App Phys Lett 1997;71:1430.
[10] Coll BF, Jaskie JE, Markham JL, Menu EP, Talin AA, vonAllmen P. Mat Res Soc 1998;498:185.
[11] deHeer WA, et al. Science 1995;270:1179.
[12] deHeer WA, et al. Adv Mats 1997;9:87.
[13] Wang OH, Corrigan TD, Dai JY, Krauss AR. App Phys Lett 1997;70:3308.
[14] Obraztsov AN et al. ISDED-2, Osaka, 1998.
[15] Bandis C, Pate BB. App Phys Lett 1996;69:366.
[16] Groning O, Kuttel O, Groning P, Schlapbach L. App Phys Lett 1997;71:2253.
[17] Robertson J. Mat Res Soc Symp Proc 1997;471:217.
[18] Robertson J. Mat Res Soc Symp Proc 1998;498:197.
[19] Ristein J, Schafer J, Ley L. Diamond Related Mats 1995;4:508.
[20] Schafer J, Ristein J, Ley L. J Vac Sci Technol A 1997;15:408.
[21] Cui JB, Ristein J, Ley L. Phys Rev Lett 1998;81:429.
[22] J Rutter M, Robertson J. Phys Rev B 1998;57:9241.
[23] Modinos A. Surface Sci 1974;42:205.
[24] Xu WS, Tzeng Y, Latham RV. J Phys D 1993;26:1776.
[25] Huang et al ZH. J Vac Sci Technol B 1995;13:522.
[26] Pate B, Chang WY, Bandis C, Okano K. ISDED-2, Osaka 1998.
[27] Collins AT. Physica B 1993;185:284.
[28] Davis RF. J Vac Sci Technol A 1993;1:829.

Pergamon

Carbon 37 (1999) 765–769

CARBON

Insights on the deposition mechanism of sputtered amorphous carbon films

S. Logothetidis*, M. Gioti, P. Patsalas, C. Charitidis

Department of Physics, Aristotle University of Thessaloniki, 54006, Thessaloniki, Greece

Received 17 June 1998; accepted 3 October 1998

Abstract

Low energy Ar^+ ion bombardment (LEIB) during growth of amorphous carbon (a-C) films deposited with magnetron sputtering (MS), results to dense films, rich in sp^3 C–C bonds, and exhibit high hardness and compressive stress. We present here a preliminary study of the growth mechanism of a-C films deposited with negative bias voltage (LEIB) in terms of their composition, density and mechanical properties. The experimental results showed that stress and hardness are directly related with the sp^3 C–C bonding in the film and described well with the so far proposed models on the formation mechanism of tetrahedral carbon. However, the film density, that is a composite property, was found to depend not only on the sp^2 and sp^3 content but also on a new, denser than graphite, carbon phase when the Ar^+ ion energy is above ~130 eV. © 1999 Elsevier Science Ltd. All rights reserved.

Keywords: A. Amorphous carbon; B. Plasma sputtering; D. Elastic properties; Microstructure

1. Introduction

Amorphous hydrogen-free carbon (a-C) films rich in sp^3 sites have a rapidly increasing area of applications in improving nanoscale components such as those used in magnetic hard discs, wear-resistant coatings, and they are promising in semiconductor devices and optical film applications [1]. a-C films with a high fraction of sp^3 bonded carbon (ta-C) have been deposited with energetic carbon ions, C^+, produced by several techniques such as filtered cathodic vacuum arc (FCVA) [2–4], mass selected ion beam (MSIBD) [5,6], laser ablation [7,8] and dual ion beam [9]. The above techniques provide well-defined deposition conditions and as a result the deposited species and their energies are known and controllable. Recently, it has been reported that dense and highly tetrahedral amorphous carbon films may also be deposited by magnetron sputtering (MS) with intense ion plating (MS/IP) [10,11], unbalanced MS [12] and negatively substrate biased MS [13–15].

The statistical character of the sputtering technique and the use of heavy Ar^+ ions may modify the subplantation

process and the relaxation mechanism that explains the growth of ta-C films when energetic C^+ species are used. In sputter process, the ion bombardment can be achieved by applying a negative bias voltage V_b to the substrate [13,14]. During growth the film is bombarded with ions that penetrate the surface providing most of their kinetic energy (E_i), that is in first approximation the sum of the average energy E_a of the discharge plus the one provided by the V_b ($E_i = e|V_b| + E_a$), to the growing film in the form of thermal spikes [16,17]. These spikes induce local rearrangements of the film's near surface atoms [18] resulting in very dense or even new metastable structures rich in sp^3 bonds. High density and compressive stress are the main characteristics of ta-C films and it has been proposed that the densification controls the sp^3 C–C bonding and the stress is arising from it (densification model) [19,20], in contradiction with the compressive stress model [2,10], where the stress is the causative factor.

We present here an attempt to investigate whether the considerations of the so far proposed models for the development of a-C films by ionized carbon species can be applied in the case of sputter deposited a-C films as well. We also present a comparative study of the compositional, structural and mechanical properties of sputtered a-C thin films with respect to the bias voltage applied to the substrate during deposition. We have found that there are

*Corresponding author. Tel.: +30 31 998174; fax: +30 31 246484.
E-mail address: logot@ccf.auth.gr (S. Logothetidis)

three regimes of V_b values where the deposition mechanism changes significantly. Moreover, the Ar^+ ion bombardment with energy above 130 eV ($V_b = -100$ V) seems to lead to the formation of a new carbon phase in a-C films.

2. Experimental details

The a-C films were deposited by rf magnetron sputtering, using a 6 in. graphite target (99.999% purity) in a deposition chamber with a base pressure better than 1×10^{-7} mbar. The c-Si (100) substrates were sputter-cleaned for 5 min by Ar^+ ion bombardment with very low energy. The substrates were located 65 mm above the target, and coated using a sputtering power of 100 W and an Ar working pressure of 2×10^{-2} mbar. All the films were deposited at room temperature (RT) in one layer (films A) or in sequential thin layers (films B) to the same total thickness of about ~300 Å. The energy and the flux of the ions (mainly Ar^+) reaching at the growing film surface was varied by applying an external V_b to the substrate in the range from $+10$ to -200 V. The above experimental process conditions correspond to the optimum one in order to obtain a-C films, even with positive V_b, with sp^3 content above 30%.

In-situ spectroscopic ellipsometry (SE) spectra were obtained (with a phase modulated ellipsometer attached on the deposition system) in the energy region 1.5–5.5 eV after each layer was deposited and upon completion of film growth. The internal stresses were derived by cantilever laser beam technique measurements of Si substrates' curvature, before and after film deposition, using Stoney's formula [14]. The film densities were obtained by X-ray Reflectivity (XRR) in a SIEMENS D-5000 diffractometer equipped with a Goebel mirror and a special Reflectivity sample stage. The XRR measurements were performed in the form of θ–2θ scan for incidence angle 0–3.2° with 0.0025° step and described in detail elsewhere [21]. The hardness measurements were conducted by the Continuous Stiffness Measurements technique in a Nano Indenter XP system with a Berkovich, three-sided pyramid diamond indenter. The system has load (displacement) resolution of 50 nN (<0.01 nm). Experiments were performed in a clean-air environment of ~45% relative humidity and 22(1)°C ambient temperature.

3. Results and discussion

In sputtering process many deposition parameters may vary in order to have the optimum conditions for the deposition of highly sp^3 bonded a-C films. However, it seems that the most important parameters are the substrate temperature (T_s) and the substrate bias, V_b.

In order to investigate the effect of T_s on the film

density and composition we deposited the a-C films with two different procedures. The first involved deposition in one layer (films A) and the other was deposition in thin sequential layers (films B). The variation of density r with the film thickness is shown in Fig. 1. The solid (dotted) curve corresponds to films A (B) deposited with $V_b = -20$ V. Both films A and B show a rapid increase in density after the initial stages of growth suggesting that the ion bombardment results in a closer packed material with less microvoids and promotes the formation of sp^3 sites. Furthermore, while the density of films B increases with thickness, with a tendency to saturate above ~70 Å, films A exhibit a reduction in density for thickness above 100 Å. This is because the bombardment with energetic Ar^+ ions during the film growth arises the temperature of the actively growing surface region [6,18]. Thus, in films B the deposition temperature is kept close to RT whereas in films A for continuous deposition, e.g. longer than 6 min (~100 Å thick), a T_s above 50°C was measured on the substrate. When the thermal energy transferred to the surface is more than the bonding energy, a reconstruction is observed that favors the formation of stable phases of carbon, i.e. the formation of sp^2 bonds [22]. This relaxation temperature of 50°C is lower than the reported one by Chhowalla et al. for FCVA deposited a-C films [23].

In-situ SE measurements were used to determine the volume fractions of sp^3 and sp^2 contents. Pseudodielectric function spectra $\langle \varepsilon(\omega) \rangle$ were obtained after the deposition of several ~300 Å thick a-C films, grown in sequential layers, in the energy region 1.5–5.5 eV. Assuming that the deposited film of thickness d is a composite material (consisting of sp^2 [24], sp^3 [8] components and voids), the measured $\langle \varepsilon(\omega) \rangle$ was analyzed by means of the Bruggeman Effective Medium Theory (BEMT) in combination with the three-phase (air/a-C/Si substrate) model [24]. It is important to note that the absolute values of sp^2 and sp^3

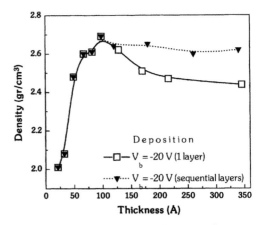

Fig. 1. Variation of a-C films density vs thickness. The films grown with negative V_b in one layer and in sequential layers.

content may vary depending on the reference dielectric functions used in the BEMT as it was discussed elsewhere [13,14].

Fig. 2 shows the variation in sp³ volume fraction versus V_b. The dependence of sp³ content on bias (ion energy) is found to be similar with that found in films deposited by other deposition techniques [3,10,23] (but with lower absolute values) and follow rather well the theoretical considerations of Robertson [18] and Davis [20]. For V_b>0 V (Regime I) there is no Ar⁺ ion bombardment but only Ar neutrals bombard the film surface. Thus, the C species are deposited softly onto the substrate in near equilibrium conditions that result in the formation of a stable carbon phase, i.e. sp² bonded, and a small amount of sp³ bonded material. For −100 V<V_b<0 V (Regime II, low energy ion bombardment), there is a plateau with high sp³ content that supports the existence of a subplantation mechanism. This subplantation mechanism may not be direct, since Ar are not the depositing species, but an indirect one i.e. the Ar⁺ ions transfer their energy to the surface C atoms and these are subplanted below the surface. For V_b<−100 V (Regime III, ion energy above 130 eV) we observe in Fig. 2 a reduction in sp³ content similar to the so far reported [3,10,23] suggesting that the thermalization mechanism of the spike model is valid also in sputter deposited a-C films. However, this argument is not confirmed by the density results in Regime III as it will be discussed below.

The dependence of compressive stress and hardness on the V_b for a-C films 300 Å thick are also shown in Figs. 3 and 4, respectively. Both of them exhibit the same dependence with the sp³ content on V_b (Fig. 2). There is a sharp increase in going from ground to negative V_b and a broad non-symmetrical peak at V_b=−40V. Both the stress relaxation and hardness above V_b=−100V are concurrent

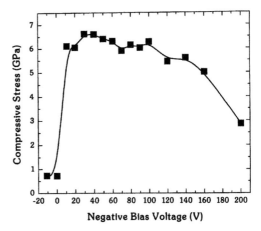

Fig. 3. The variation of compressive stress of sputtered a-C films, ~300Å thick, versus V_b.

with the decrease of the sp³ content. The stress behavior can be described successfully by Davis' formula [20] and will be discussed elsewhere.

If we take into account that the hardness of a solid is defined by the product of binding energy E_b and the covalency a_c divided by the interatomic spacing (or bond length d) [25] $H \sim (E_b a_c)/d^2$, we can conclude that in pure covalent bonds (a_c=1), a hard a-C film must have a short bond length, which corresponds to a highly sp³ bonded material. On the other hand, an a-C film rich in sp² sites must be a soft material due to the layered structure and the weak interplanar bonding. That is, the results of hardness in Fig. 4 match nicely with those of sp³ content, in Fig. 2,

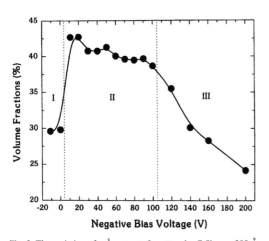

Fig. 2. The variation of sp³ content of sputtered a-C films, ~300 Å thick, versus V_b.

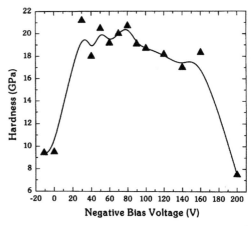

Fig. 4. The variation of hardness of sputtered a-C films, ~300 Å thick, versus V_b.

versus V_b or energy of Ar^+ ion bombarding the film during deposition.

Finally, we studied the dependence of film density on V_b. Our results are based on both SIMPLEX and MONTE-CARLO fittings that are in excellent agreement with the experimental results as well as to each other. The density values were calculated by analyzing the XRR data with SIEMENS/REFSIM software using both Simplex and Monte-Carlo fitting procedures assuming either a single-layer (a-C film/Si-substrate) or a two-layer (a-C film/SiC layer/Si) model and are presented (solid squares) in Fig. 5. The former model was mainly used in films belonging to Regime I and II, whereas the latter fits better the spectra obtained from films of Regime III. In a number of previous studies, the density was used to calculate the sp^2 and sp^3 content in a-C films by comparing it to those of bulk crystalline graphite and diamond (2.25 and 3.51 g/cm^3, respectively). The density in thin films, either nanocrystalline or amorphous, can be less compared to their bulk counterpart due to several reasons. In a-C films, however, this difference could be even larger or may occur the opposite since new forms of carbon, denser than graphite, may exist [26–28] providing wrong estimation of sp^2/sp^3 ratio.

The density, using the sp^2 and sp^3 volume fractions calculated by SE, was also calculated in first approximation through the following expression:

$$\rho = \rho_{sp^2} \cdot f_{sp^2} + \rho_{sp^3} \cdot f_{sp^3}, \tag{1}$$

where $\rho_{sp^2}(\rho_{sp^3}) = 1.88(3.5)$ g/cm^3 is the reference density and $f_{sp^2}(f_{sp^3})$ is the volume fraction of sp^2 (sp^3) bonded material. A comparison of these density values (open cycles) with the directly measured density by XRR (solid squares) is shown in Fig. 5. The agreement between the measured (XRR) and calculated (SE) density is used to validate the SE results regarding the sp^2 and sp^3 volume fractions. Regime I, in Fig. 5, is characterized by low density values ($\sim$1.88 g/cm^3) in agreement with the low sp^3 content as it was shown in Fig. 2. In Regime II the density is almost constant, $\sim$2.62 g/cm^3, and independent from V_b.

For $V_b < -100$ V ($E > 130$ eV, Regime III) the voids content becomes considerably low due to the microvoids collapsing under ion bombardment and besides a new denser than sp^2 bonded amorphous carbon phase is formed. The reduction of sp^3 content and the increase in density due to formation of the new carbon phase suggest that there is no linear relation between these two quantities as proposed by Robertson [19] and Davis [20]. Density is a composite property and as long as a material consists of several phases is an average of all phases. For example, if a new, denser than graphite, carbon phase is formed in a-C films then the film density will also depend on its content and the relative ratio to sp^3 phase. Thus, the model proposed by Robertson may fail in Regime III because it does not take into account the formation of this dense carbon phase and suggests that closed packed carbon is necessarily equivalent to sp^3 formation.

The sputter deposited a-C films may have different behavior since this process has a statistical character. The deposition species are mainly C neutrals and a few C^+ ions and involves energetic Ar species (neutrals and/or ions). Furthermore, the deposition process includes different kinds of species (C, C^+, Ar, Ar^+) and in combination with the fact that Ar species are bigger and heavier than C^+ ions could modify accordingly the proposed subplantation process and the relaxation mechanism.

The main benefit in applying negative V_b during deposition is the bombardment of the growing film with controllable flux (Φ_i) and kinetic energy (E_i) of Ar^+ ions. The ratio of Φ_i to the carbon atom flux, Φ_i/Φ_c, being about 1/3 in our case, controls together with E_i the film properties. The absolute values of internal stress and density in our sputtered films presented here are significantly higher comparing to those reported by E. Mounier et al. [12], and deposited with both conventional and unbalanced MS and lower than those reported by J. Schwan et al [11] and deposited by MSIP. This is because their Φ_i/Φ_c is about 0.2 and 5.4, respectively, whereas in our case $\Phi_i/\Phi_c(\sim 0.3)$ is between them.

The increase of density $\Delta\rho$ because of the particle bombardment was calculated, for the case of energetic C^+ species, by Robertson [16,17] and is given by the following formula:

$$\Delta\rho = \frac{\rho_0 \cdot f}{\Phi_c/\Phi_i - f + 0.016p \cdot (E/E_0)^{5/3}}, \tag{2}$$

where ρ_0 is the density when the particle energy is below a

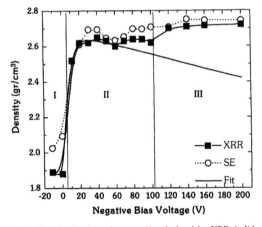

Fig. 5. The density dependence on V_b calculated by XRR (solid squares) and SE (open cycles). The prediction of the densification model is also shown (solid line).

critical one, Φ_c is the carbon species flux, Φ_i is the ion flux, p is of order 1, $f = 1 - \exp[(E_p - E)/E_{sp}]$ is the ion penetration probability, E_p is the penetration threshold, E_{sp} is a spread parameter and E_o is the activation energy for atomic diffusion. The solid curve in Fig. 3 describes an attempt to fit the density obtained by XRR up to $V_b = -60$ V ($E = 90$ eV) using Eq. (2). The best fit parameters are: $E_p = 30$, $E_{sp} = 5.8$ and $E_o = 16$ eV and $\Phi_i/\Phi_c = 0.3$. Similar values (not presented here) were obtained by fitting the compressive stress values versus V_b up to -200 V. The model follows rather well the experimental stress results but fails to describe the density of a-C films produced with Ar^+ energies above ~100 eV. This is because the above phenomenological model was proposed for the subplantation and deposition of energetic C^+ species. On the other hand, in sputter deposition many different species are involved that they have different sticking coefficients and ionization fractions and new carbon phases may be formed. Additionally, in the sputtered a-C films, Ar can be also included in the films as an impurity or defect. A large amount of Ar could produce a strain field in the film resulting to high compressive stresses. This is a common feature for all sputtered films. An attempt to account it has been reported for a-C films by Schwan et al [11] and Mounier et al [12]. They found in sputtered a-C films an Ar concentration ~6 and 2%, respectively. That is, besides that Ar species may participate in the development of the a-C films in two different subplantation processes: i) the direct one (Ar species with energy above ~30 eV are subplanted below the surface), and ii) the indirect one (Ar species transfer their energy to the surface C atoms that are then subplanted), Ar could also contribute as an impurity to the sputter process. Thus, the internal structure and properties of the sputtered a-C films, discussed above, either support that the participation of Ar modifies the knock on implantation mechanism and makes the thermal relaxation process for the formation of sp^3 bonding to appear more complicated or that the mechanism needs to improve further in order to account for the formation of additional metastable phases.

4. Conclusions

The variation of the composition, compressive stress, hardness and density of sputtered a-C films was studied as a function of substrate bias voltage (Ar^+ ion energy). It was found that the internal stress and hardness are directly related with the sp^3 content and exhibit the same dependence on the Ar^+ ion energy bombarding the film surface. On the other hand, the film density, which is a composite property, does not exhibit the same dependence on ion energy for $E > 130$ eV. This can be explained with the existence of a new metastable phase which is formed and favored in these energies at least in sputtering process.

Acknowledgements

This work was supported in part by EU BRPR-CT96-0265 project.

References

[1] Bhushan B, Kellock AJ, Cho NH, Ager W III. J. Mater Res 1992;7, 404.
[2] McKenzie DR, Muller D, Pailthorpe BA. Phys Rev Lett 1991;67, 773.
[3] Fallon PJ, Veerasamy VS, Davis CA, Robertson J, Amaratunga GAJ, Milne WI, Koskinen J. Phys Rev B, 1993;48, 4777.
[4] Silva SRP, Xu S, Tay BX, Tan HS, Milne WI. Appl Phys Lett 1996;69 (4), 491.
[5] Lifshitz Y, Kasi SR, Rabalais JW, Eckstein W. Phys Rev B 1990;41, 16468.
[6] Lifshitz Y, Lempert G, Grossman E. Phys Rev Lett 1997;72, 2753.
[7] Cuomo JJ, Pappas DL, Bruley J, Doyle JP, Saenger KL. J Appl Phys 1991;70, 1706.
[8] Xiong F, Wang YY, Chang RPH. Phys Rev B 1993;48, 8016.
[9] André B, Rossi F, van Veen A, Mijnarends PE, Schut H, Delplancke MP. Thin Solid Films 1994;241, 171.
[10] Schwan J, Ulrich S, Roth H, Ehrhardt H, Silva SRP, Robertson J, Samienski R, Brenn R. J Appl Phys 1996;79 (3), 1416.
[11] Schwan J, Ulrich S, Theel T, Roth H, Ehrhardt H, Becker P, Silva SRP. J Appl Phys 1997;82(12), 6024.
[12] Mounier E, Pauleau Y. Diam Real Mater 1997;6, 1182.
[13] Logothetidis S. Appl Phys Lett 1996;69, 158.
[14] Logothetidis S, Gioti M. Mater Sci Eng B, 1997;46, 119.
[15] Gioti M, Logothetidis S. Diam Relat Mater 1998;7, 444.
[16] Seitz F, Koehler JS. Sol St Phys 1956;2, 305.
[17] Müller KH. Phys Rev B 1987;35, 7906.
[18] Hofsäss H, et al. Appl Phys A, Mater Sci Processing 1998;66:153.
[19] Robertson J. Diam Rel Mat 1994;3:361.
[20] Davis CA. Thin Solid Films 1993;226:30.
[21] Logothetidis S, Stergioudis G. Appl Phys Lett 1997;71:2463.
[22] Logothetidis S, Petalas J, Ves S. J Appl Phys 1996;79:1040.
[23] Chhowalla M, Robertson J, Chen CW, Silva SRP, Davis CA, Amaratunga GAJ, Milne WI. J Appl Phys 1997;81:139.
[24] Aspnes DE. Thin Solid Films 1982;89:249.
[25] Kisly PS. In: Almond EA, Brookes CA, Warren R, editors. Proc of the Int Conf on the Science of Hard Materials, 1984.
[26] Kelires PC. Phys Rev B 1993;47:1829.
[27] Cohen ML. Phys Rev B 1991;43:6742.
[28] Amaratunga GAJ, Chhowalla M, Kielly GJ, Alexandrou I, Aharonov R, Devenish RM. Nature 1996;383:321.

Pergamon

Carbon 37 (1999) 771–775

CARBON

C–H bonding of polymer-like hydrogenated amorphous carbon films investigated by in-situ infrared ellipsometry

T. Heitz*, B. Drévillon, C. Godet, J.E. Bourée

Laboratoire de Physique des Interfaces et des Couches Minces (UMR 7647 CNRS) Ecole Polytechnique, 91128 Palaiseau, France

Received 17 June 1998; accepted 3 October 1998

Abstract

Polymer-like hydrogenated amorphous carbon films have been deposited in a microwave-assisted radiofrequency plasma reactor. Vibrational properties are investigated in-situ by infrared ellipsometry as a function of ion bombardment. By choosing either hydrogen or helium in the microwave gas mixture, films ranging from nearly purely saturated polymer to olefinic/aromatic films can be obtained. Both absorption bands and transparent regions are studied, leading to a determination of the dielectric constant as well as the vibrational frequencies, widths and intensities as a function of macroscopic properties like density and optical gap. Hydrogen atoms attached to sp^2-carbon are detected even in highly saturated films. The sensitivity of the vibrational oscillator strength to the matrix effect is used to probe the local environment of hydrogenated carbon atoms. Quantitative determination of CH_n group densities in the different configurations indicates that sp^3 CH groups are dominant and that the local structure evolves with atomic hydrogen concentration. © 1999 Elsevier Science Ltd. All rights reserved.

Keywords: A. Amorphous carbon; Carbon films; B. Plasma deposition; C. Ellipsometry; D. Microstructure

1. Introduction

Hydrogenated amorphous carbon (a-C:H) shows a wide range of outstanding physical and chemical properties [1,2] like high hardness, electrical resistivity, near infrared transparency or room temperature luminescence in the visible range: a-C:H has been used as active layers in electroluminescent displays [3]. Electro-optical and mechanical properties take their origin in the ability of carbon atoms to have mainly two sorts of hybridisation states, sp^2 (threefold coordination) or sp^3 (fourfold coordination). By varying the hydrogen concentration and sp^2/sp^3 ratio, soft polymer-like luminescent films with a 35–55% hydrogen content or hard amorphous carbon with less hydrogen (about 20%), called for historical reasons diamond-like carbon (DLC), can be obtained. The particular properties of all these different types of films are closely related to the microstructure. This study will focus on plasma-deposited carbon films that show a strong photoluminescence (PL) intensity [4] C–H bonding will be studied as a function of film density.

By increasing the ion energy impinging on the film surface, monitored by the RF power, or decreasing the atomic hydrogen concentration in the plasma, sp^2 configurations are favoured leading to more absorbent materials with a lower H fraction, from 55 to 35 at.% [5]. As more than one atom out of three is hydrogen, the microstructure can be mostly analysed through the C–H bonding. By measuring in-situ infrared (IR) ellipsometric spectra as a function of film thickness, CH_n group densities are determined quantitatively. The hydrogen redistribution is investigated going from highly saturated polymer-like to harder films and some guidelines for microstructural models can be given.

2. Experimental details

2.1. Film deposition

a-C:H films are deposited at a low pressure (300 mTorr) using a double plasma radiofrequency (RF)/microwave (MW) technique [5]. MW plasma is created in a quartz tube where a mixture of hydrogen/argon or helium/argon is decomposed. Butane is injected near the substrate holder

*Corresponding author. Tel: +331-693-332-19; Fax: +331-693-330-06.

where a RF discharge is formed. The deposition temperature is 70°C or 150°C when using H_2 or He, respectively. To obtain a better signal/noise ratio, a-C:H is deposited on palladium substrates which have been pre-treated [6] to increase film adhesion. UV–visible ellipsometry, Rutherford Backscattering (RBS) and Elastic Recoil Detection Analysis (ERDA) allow the determination of the optical gap, atomic density and hydrogen concentration.

2.2. IR measurements

Vibrational properties are investigated in-situ as a function of film thickness. In the 1000–4000 cm^{-1} range, bending (near 1400 cm^{-1}) and stretching (near 3000 cm^{-1}) C–H modes are studied using HgCdTe and InSb detectors, respectively. IR ellipsometry, based on a double modulation (optical path and light phase) technique [7], allows the analysis of weak vibrations [8]. IR properties are investigated through the (ψ, Δ) spectra defined by:

$$\frac{r_p}{r_s} = \tan \psi \, e^{i\Delta} \qquad (1)$$

where r_p and r_s are the reflection coefficients for p- or s-polarised light. (ψ, Δ) spectra are recorded with a 4 cm^{-1} step. To subtract the substrate contribution to the signal, we use the ellipsometric density D which has the expression:

$$D = [\ln \tan(\psi_s) - \ln \tan(\psi)] + i(\Delta_s - \Delta) \qquad (2)$$

(ψ_s, Δ_s) is the substrate spectrum. In the case of thin weakly IR absorbent films deposited on metallic substrates, a first order expansion of D on σd, where σ is the wavenumber and d the film thickness, leads to:

$$D = K\left[\left(1 - \frac{1}{\varepsilon_\infty}\right) + \left(\frac{1}{\varepsilon_\infty^2}\sum_k \Delta\varepsilon_k\right) \right] d\sigma \qquad (3)$$

where K is a constant. The film dielectric function is considered as a constant (ε_∞) outside the absorption bands. The contribution of the kth vibration is modeled by a Lorentzian shape:

$$\Delta\varepsilon_k = \frac{f\sigma_0^2}{\sigma_0^2 - \sigma^2 + i\Gamma\sigma} \qquad (4)$$

Γ is a damping parameter and σ_0 is the eigenfrequency. The coefficient f (expressed without units) represents the absorption intensity.

According to Eq. (3), slope analysis of the IR background allows the determination of ε_∞. After linear subtraction, vibrational bands are decomposed by using Lorentzian functions with the same set (σ_0, Γ) for all thicknesses. The complete analysis method, which is based on the study of the imaginary part of D (Im (D)) is described elsewhere [6]. Note that vibrations are identified by a negative inflection in Im (D) IR spectra.

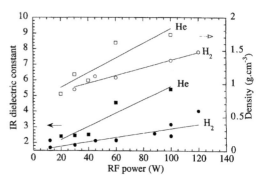

Fig. 1. Dielectric IR constant ε_∞ (solid symbol) and density (open symbol) as a function of radiofrequency (RF) power. He (squares) or H_2 (circles) gas is mixed with argon and decomposed in the microwave plasma.

3. Vibrational analysis

3.1. Out-of-band region

In Fig. 1, ε_∞ and density are plotted versus RF power for both types of MW gas mixtures. A similar increase with P_{RF} can be observed and is related to the global increase of C=C unit concentration. But C_4H_{10}/He/Ar mixture makes it possible to obtain denser films with a higher dielectric constant. ε_∞ is in fact due to the contribution of π–π^* electronic transitions and depends on the product of the optical transition matrix element and the joint density of states. UV–visible measurements [5,9,10] have confirmed that the latter product increases with ion energy. Simultaneously, the optical gap is found to decrease from 3 to 0.5 eV when films become denser [5].

3.2. Bending region

As shown in Fig. 2, the C–H bending region has been decomposed using six vibrations located at: 1370–1380, 1404, 1418, 1437, 1450 and 1465 cm^{-1}. For all films, the line positions are constant within 5 cm^{-1} except for the 1370–1380 cm^{-1} band, attributed to symmetric (sym) CH_3 groups. According to hydrocarbon IR spectra [11], the methyl bending frequency is sensitive to the presence of π bonds as first neighbours. Symmetric and antisymmetric (asym) methyl line positions are proved to be located at 1379 and 1458–1467 cm^{-1} for C–CH_3 units and at 1370 and 1440 cm^{-1} for =C–CH_3 groups. Using IR handbooks, band positions listed above can be identified (see Table 1) and compared with other authors' data [12]: vibrations at 1404 and 1418 cm^{-1}, observed even in highly hydrogenated films, clearly evidence the presence of π bonds in olefinic (olef) configuration proving that a purely sp^3 network cannot be obtained, even at low ion energy. Our sp^2 CH_n frequency observations do not correspond to

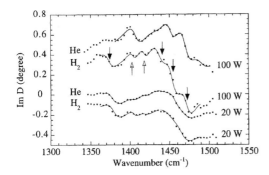

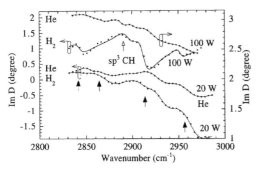

Fig. 2. Im (D) spectrum of a-C:H films in the bending (a) or stretching (b) region. He or H_2 is decomposed in the microwave plasma. Different RF powers have been used.

Dischler's attributions [12]. If the 1437 cm^{-1} peak were attributed to sp^2 CH in aromatic configuration (arom), a clear band should be observed in the stretching region for almost all films, which is not the case as will be detailed further.

3.3. Stretching region

sp^3 CH$_n$ band stretching frequencies range from 2800 to 3000 cm^{-1}. The CH$_2$ and CH$_3$ bands are generally sufficiently strong to be detected. As shown in Fig. 2(b), another inflection located between sp^3 CH$_3$ (sym) and sp^3

CH$_2$ (asym) vibrations (see Table 1) can also be observed and is attributed to sp^3 CH groups. This configuration is only directly detected in dense films. For highly hydrogenated materials, decomposition procedures using five Lorentzian functions are necessary to separate the IR absorption of each group. The sp^3 CH line frequency is proved to decrease when ion bombardment increases.

sp^2 CH$_n$ vibrations, which are expected in the 3000–3100 cm^{-1} range, show extremely weak absorption intensities in transparent films. In more absorbent films, inflections, reported in Table 1, are detected corresponding to both olefinic and aromatic configurations. The bandwidths remain always low (between 15 and 25 cm^{-1}) compared with sp^3 CH$_n$ ($n=2,3$) peaks (from 25 to 50 cm^{-1}).

4. Effective charges

Let us recall that the oscillator strength f deduced from band decomposition is related to the oscillator density N by the relation:

$$f = A \frac{Ne*^2 f_{int}^2}{\sigma_0^2} \tag{5}$$

where A is a constant. f_{int} is the local field factor [13] correction: in our study, the local field is supposed to be described by the Lorentz approximation [14]. Using Eq. (5), effective charges are proved to be related to absorption intensity ratios between bending/stretching and (sym)/(asym) modes [6]. As far as sp^3 CH$_2$ and sp^3 CH$_3$ are concerned, the latter ratios can be fairly simulated assuming constant effective charges, similar to those of hydrocarbon compounds [11]. The type of carbon hybridisation of sp^3 CH$_n$ ($n=2,3$) neighbours must however be taken into account, as for line positions.

The effective charge of sp^3 CH is determined by subtracting the concentration of hydrogen atoms involved in methyl and methylene groups from the total hydrogen concentration deduced from ERDA. This method assumes that hydrogen is preferentially bonded to sp^3 C, more exactly that sp^2 CH$_n$ group concentration is negligible compared with that of sp^3 CH units. One also supposes that almost every H atom is bounded to C, which is

Table 1
Line positions determined through band decomposition of bending and stretching regions

Bending sp^3 and sp^2	1370–1380 CH$_3$ (sym)	1404 sp^2 CH olef.	1418 sp^2 CH$_2$	1437 =C–CH$_3$ (asym)	1450 sp^3 CH$_2$	1467 –C–CH$_3$ (asym)
Stretching sp^3	2845–2860 CH$_2$ (sym)	2870 CH$_3$ (sym)	2900–2915 CH	2915–2930 CH$_2$ (asym)	2960 CH$_3$ (asym)	
Stretching sp^2	2975–2980 CH$_2$ (sym)	3014 CH olef.cis	3035 CH olef.trans	3050 CH arom.	3065–3080 CH$_2$ (asym)	

Attributions are consistent with hydrocarbon infrared spectra.

consistent with other studies [15]. The sp^3 CH effective charge is found to strongly decrease when the RF power increases, from 0.1 to 0.02 (in e unit) which can be related to the formation of distorted weakly hydrogenated C network.

5. Hydrogen distribution

Thanks to effective charge calculations, CH_n group densities can be evaluated in a quantitative way. Densities of H atoms involved in sp^3 CH_n groups are plotted in Fig. 3 versus film density, showing that sp^3 CH is by far the majority group. Methyl and methylene group concentrations monotonously decrease versus density which is interpreted as a cross-linking of the polymeric phase. The sp^3 CH group density increases from $1-4 \times 10^{22}$ cm^{-3}. For low density films, H is nearly equally distributed in CH, CH_2 and CH_3 configurations.

To evaluate sp^2 CH_n group concentration, effective charges of hydrocarbons have been considered. Bending vibrations located at 1404 and 1418 cm^{-1} allow the calculation of olefinic hydrogenated configuration densities. In the stretching mode, the determination of the absorption intensity for each band is not straightforward: the 3000–3100 cm^{-1} region cannot be properly decomposed using Lorentzian shapes. Consequently, the ratio between inflection amplitudes of 3050 cm^{-1} (arom. CH) and 3014–3035 cm^{-1} (olef. CH) bands is used to determine the relative concentration of aromatic configurations, assuming the same effective charges as in hydrocarbons [11]. In Fig. 4, the hydrogen distribution in olefinic or aromatic configurations is plotted against ε_∞. From $\varepsilon_\infty = 2$ to 3.5 (that is for a density from 0.85 to 1.2

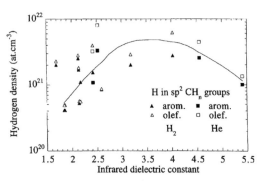

Fig. 4. Oscillator densities of sp^2 CH_n olefinic groups as a function of infrared dielectric constant (ε_∞). H concentration attached to sp^2 olefinic (open symbols) or aromatic (solid symbols) configurations is plotted. Films have been obtained using Ar/C_4H_{10} mixed with He (squares) or H_2 (triangles).

g cm^{-3}), the sp^2 CH_n group concentration increases to 2×10^{21} cm^{-3}. A decrease is then observed as ε_∞ and film density reach higher values.

6. Discussion and conclusion

As far as the saturated carbon phase is concerned, by comparing –C–CH_3 and =C–CH_3 band intensities, vibrational analysis proves that the neighbourhood of methyl groups evolves as a function of increasing density from a purely sp^3 to a mixed sp^2/sp^3 configured environment in ratio 1: 2. Therefore, sp^2 and sp^3 network are thought to be closely interconnected with each other. Moreover, as CH_2 groups are in a minority, the saturated polymeric phase does not consist in $(CH_2)_n$ linear chains. The increase of

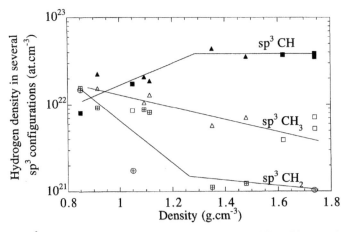

Fig. 3. Oscillator densities of sp^3 CH_n units. Films have been obtained using Ar/C_4H_{10} and He (solid squares for CH), (open squares for CH_3), (surrounded cross for CH_2) or H_2 (solid triangles for CH), (open triangles for CH_3), (cross with open squares for CH_2).

sp^3 CH groups versus film density is consistent with cross-linking. However, in dense films, end group concentration (i.e. CH_3) is five times lower than for single CH: ramifications are numerous but do not always end by a methyl group. Some three-dimensional complex sp^3 arrangements may then be formed under energetic ion flux: effective charge of hydrogen atoms bonded to such structures is expected to be low [16]. The sp^3 CH frequency decrease could also be an indication of such structure formation.

Low vibrational widths of sp^2 CH_n groups may suggest a higher local order in sp^2-rich environment. Using a geometrical model, Angus has argued [17] that strain energy is released when π bonds are formed. The optical gap of dense films is around 0.5 eV [5] whereas the ratio of hydrogenated carbon atoms sp^2 to sp^3 is between 0.005 and 0.03. On the one hand, it should be noted that effective charges of hydrocarbons may not be valid in a-C:H films: sp^2 CH bond density may thus be underestimated. On the other hand, the low gap value could be explained by the presence of non-hydrogenated π bonds in conjugated chains.

As a conclusion, an exhaustive study of C–H bonding has been performed on plasma-deposited a-C:H films thanks to IR in-situ ellipsometry. Through measurements as function of thickness and band decomposition, densities of sp^2 and sp^3 CH_n groups have been quantitatively evaluated proving that monohydrogenated sp^3 CH is the majority unit. H distribution analysis indicates that a-C:H film structure may be modeled by a multiphase heterogeneous saturated/unsaturated twisted polymeric network including weakly hydrogenated distorted structures.

References

[1] Robertson J, O'Reilly EP. Phys Rev B 1987;35:2946.

[2] Kelires PC. Phys Rev B 1993;47:1829.

[3] Hamakawa Y, Toyama T, Okamoyo H. J Non-Cryst Solids 1989;115:180.

[4] Heitz T, Godet C, Bourée JE, Drévillon B, Chu V, Conde JP, Clerc C. In: Parsons G, Fahlen TS, Morozumi S, Seager C, Tsai CC, editors, Materials Research Society Symposium Proceedings, Vol. 508, San Francisco, 1998.

[5] Godet C, Heitz T, Drévillon B, Bourée JE., J Appl Phys 1998;84:3919.

[6] Heitz T, Drévillon B, Godet C, Bourée JE. Phys. Rev. B 1998; to be published.

[7] Canillas A, Bertran E, Andujar JL, Drévillon B. J Appl Phys 1990;68:2752.

[8] Heitz T. PhD Thesis 1998.

[9] Demichelis F, Pirri CF, Tagliaferro A. Phys Rev B 1992;45:14364.

[10] Koidl P, Wild Ch, Dischler B, Wagner J, Ramsteiner M. Materials science forum, vols. 52,53. Switzerland: Transtech Publications, 1989:41.

[11] Wexler AS. Applied spectroscopy reviews. 1967:29-98.

[12] Dischler B. Amorphous hydrogenated carbon films. In: Koidl P, Oelhaven P, editors. E-MRS Symposium Proceedinigs, vol. 17. Paris: Les Editions de Physique, 1987:189.

[13] Ratajczak H, Orville-Thomas WJ. Trans Faraday Soc 1965;61:2603.

[14] Polo SR, Wilson MK. J Chem Phys 1953;23:2376.

[15] Honeybone PJR, Newport RJ, Walters JK, Howells WS, Tomkinson J. Phys Rev B 1994;50:839.

[16] Burns W, Mc Kervey MA, Mitchell TR, Rooney JJ. J Am Chem Soc 1978;100:906.

[17] Angus JC, Jansen F. J Vac Sci Technol A 1998;6:1778.

Pergamon

Carbon 37 (1999) 777–780

CARBON

A study of the effects of nitrogen incorporation and annealing on the properties of hydrogenated amorphous carbon films

R.U.A. Khan, A.P. Burden, S.R.P. Silva*, J.M. Shannon, B.J. Sealy

School of Electronic Engineering, Information Technology and Mathematics, University of Surrey, Guildford, Surrey, GU2 5XH, UK

Received 16 June 1998; accepted 3 October 1998

Abstract

The electronic properties of hydrogenated amorphous carbon films deposited using a Plasma Technology DP800 radio frequency plasma-enhanced chemical vapour deposition system are investigated. Films deposited on the driven electrode have a Tauc optical band-gap of 0.9–1.2 eV, a refractive index of 1.8–2.3, and are hard and diamond-like. However, films deposited on the earthed electrode are softer and polymer-like with a Tauc optical band-gap of 2–3 eV and a refractive index of 1.5–1.7. Both types of film have been grown with varying amounts of nitrogen in an attempt to dope them and measure their characteristics. Films grown on the driven electrode showed current versus voltage (I/V) characteristics indicative of Poole-Frenkel type conduction. However, the I/V characteristics of the films grown on the earthed electrode exhibited high resistivity (typically 10^{14}–10^{15} Ωcm). Thermal annealing of the films grown on the earthed electrode has also been investigated. The films containing nitrogen were found to be more sensitive to annealing. © 1999 Elsevier Science Ltd. All rights reserved.

Keywords: A. Amorphous carbon; B. Plasma deposition; Chemical vapour deposition; Annealing; D. Electrical properties

1. Introduction

Hydrogenated amorphous carbon (a-C:H) films have attracted considerable interest over recent years, as in particular they can be grown by radio frequency plasma enhanced chemical vapour deposition (rf-PECVD), allowing coverage over large areas at room temperature. Also this technique has the benefit of being well-established in the electronics industry for the deposition of amorphous hydrogenated silicon deposition [1]. Other advantages include the ability to tailor the deposited material, including the optical band gap in the range of 1 to 4 eV, and the resistivity from 10^6 to 10^{15} Ωcm. This is attributed to the ability of carbon to hybridise to form sp^2 bonds, as in graphite, sp^3 bonds, present in diamond and some hydrocarbons, and also sp^1 bonds which exist in molecules such as acetylene [2]. An increase in the sp^2 content of a-C:H causes a lowering of the optical band gap. This is understood to be due to the π states lying closer to the centre of the band gap than the σ states, and becoming more prominent as graphitisation progresses. A narrowing

of the band gap would also result in films with a higher electrical conductivity.

The effect of adding nitrogen to the plasma in order to form a-C:H:N has been widely studied; in particular it has been argued that nitrogen can act as a weak donor and electronically dope the films n-type [3]. Other effects that have been noted include the passivation of electronic defects [3] and stress relief in the films [4]. Another consequence of nitrogenation is an increase in the sp^2 content, through the promotion of graphitisation [5]. Therefore, this would also result in a narrowing of the band gap and an increase in film conductivity, so it cannot be assumed that an increase in film conductivity is solely due to the electronic doping of the material unless all these different properties are monitored.

The fact that PECVD films are formed by the glow decomposition of a hydrocarbon gaseous precursor such as methane results in a material with a high (up to 60%) hydrogen content [6]. It has been suggested that hydrogen terminates carbon dangling bonds, helping to reduce the defect density of the film to about 10^{18} cm^{-3} [7,8].

In the case of amorphous hydrogenated silicon films annealing also reduces the density of dangling bond defects leading to a better electronic material [1]. How-

---

*Corresponding author.
E-mail address: s.silva@ee.surrey.ac.uk. (S. Silva)

ever, annealing a-C:H and a-C:H:N is complicated by structural changes; in particular an increase in the sp^2 content which would ultimately lead to film graphitisation. Once again this would cause a decrease in the band gap and an increase in the electrical conductivity.

2. Experimental

The films studied in this paper were grown using a standard Plasma Technology DP800 rf-PECVD system with a capacitively coupled 13.56 MHz power source. The system was configured with a driven upper electrode and an earthed substrate table. This allowed films to be grown by attaching the substrates onto either of the water cooled electrodes. The substrates used were n- and p-type silicon of resistivity 0.01–0.02 Ωcm, Corning 7059 glass, and chromium nitride sputter-coated glass (a metallic thin film). Depositions were carried out at ambient temperature, a pressure of 200 mTorr and a rf power of 200 W. The precursor gases were methane and helium with flows of 30:75 sccm. Nitrogen was added to the plasma with a flow rate of 0–15 sccm in order to form a-C:H:N. Devices were fabricated from the carbon-on-metal (CrN) structures, by evaporating gold top contacts directly onto the films.

The films were measured using a Camspec M330 uv-visible spectrophotometer, a Sloan Dektak IIA profilometer and a Plasmos SD200 ellipsometer. Current versus voltage (I/V) measurements were carried out using a Keithley 487 picoammeter/voltage source. An 8 kVA optical rapid thermal annealer with a flowing nitrogen atmosphere was used for post-deposition processing of the samples.

3. Results and discussion

3.1. Nitrogenation of earthed electrode films

The optical results for the films as a function of nitrogen flow rate are shown in Fig. 1. It can be seen from the graph that the Tauc optical band gap reduces as the nitrogen flow rate increases. This may indicate that nitrogen promotes graphitisation by the forming of sp^2 bonds within the film. However, this would generally coincide with a corresponding increase in refractive index, which was not observed in this case. A possible explanation can be found using the Penn model [9] which relates the band gap energy (E_g) and the refractive index of a material (n) in terms of the plasmon energy E_p;

$$n^2 - 1 = (E_p/E_g)^2 \qquad (1)$$

A change in the band gap without a corresponding change in the refractive index of a material indicates a change in the plasmon energy, i.e. a corresponding decrease in the plasmon energy. As the plasmon energy is influenced by

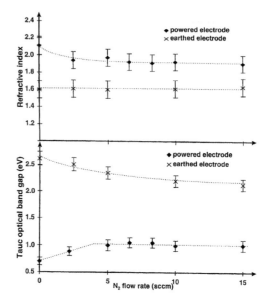

Fig. 1. Variation of refractive index and Tauc optical band gap with nitrogen flow rate into the chamber.

the joint density of states (JDOS) of the material as well as its valence density, this provides tentative evidence for a modification of states within the band gap as a result of nitrogen incorporation. In a previous study of nitrogenation [3] there was little evidence of a change in the density of films with increasing nitrogen contents. Preliminary RBS results suggest that the films grown contain up to 12% atomic nitrogen.

The current density versus applied field (J/E) plots of films grown on CrN substrates are shown in Fig. 2(b). The characteristics are symmetrical, and show only a small current density rise with applied electric field. The magnitude of the current is also very low. Yet if a J versus $E^{1/2}$ curve is plotted, straight line regions are observed with the slopes giving rise to dielectric values that are not realistic if Poole-Frenkel type conduction is assumed. Since a Schottky barrier too would give rise to a J versus $E^{1/2}$ relationship it is possible that the conduction arises due to a back-to-back Schottky barrier [7]. It is also observed that the leakage current increases as a function of the nitrogen flow rate used during film deposition. This may be correlated to a reduction of the band gap (or a movement of the Schottky barrier height) or the proposed modification of the JDOS.

3.2. Nitrogenation of powered electrode films

The effects of adding nitrogen to the plasma are shown in Figs. 1 and 2 (a). The J/E characteristics shown in Fig. 2 contrast with those derived from the films grown on the

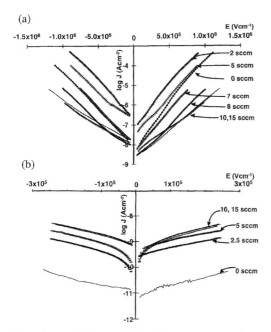

Fig. 2. *J* versus *E* characteristics for films grown under different nitrogen flow rates: a) driven electrode grown, b) earthed electrode grown.

earthed electrode in that they possess an appreciable turn-on slope. This may be analysed using the Poole-Frenkel equation [10], where σ is the conductivity, σ_0 is the prefactor which has been shown to depend on the defect density of the material, and β is a constant, depending on the absolute temperature (T) the electronic charge (q), the permittivity of free space (ϵ_0), the dielectric constant (ϵ_r) and the film thickness (d).

$$\sigma = \sigma_0 \exp[\beta V^{1/2}]; \quad \beta = \frac{q}{kT}\sqrt{\frac{q}{\pi\varepsilon_0\varepsilon_r d}} \qquad (2)$$

From plotting *ln* (current density) versus (applied electric field)$^{1/2}$ values of dielectric constant ε_r have been calculated; the values obtained being in the range of 4–5, closely fitting the square of the refractive index. Hence, these films appear to exhibit bulk-limited Poole-Frenkel conduction. The same analysis does not yield acceptable values of ε_r for the films grown on the earthed electrode, where the expected value of ε_r would be 2–3 but the derived value is in the range of 100–200.

The clearest trend in the optical variation is the initial decrease in refractive index up to a flow rate of 5 sccm nitrogen. A concomitant increase in the optical Tauc gap suggests that the films are not re-ordering into a sp^2 rich network, but possibly clearing up localised mid gap tail states that will give rise to leakage currents via hopping.

This however is apparently contradicted by the *J/E* data which suggests that the addition of a small flow of nitrogen (~2 sccm) has the effect of increasing the conductivity. Therefore, it leaves the possibility that the passivation of these defects occurs in such a manner that there is a doping effect either by substitution or by a positive N^+/C^- defect pair [3].

However, at higher flow rates of nitrogen, the conductivity appears to decrease. This more likely to be the result of the passivation of dangling bonds, causing a reduction in the Poole-Frenkel prefactor σ_0.

3.3. Annealing of earthed electrode films

The optical results for two films grown on the earthed electrode are shown in Fig. 3. In both cases the refractive index increases and the optical band gap shows a corresponding decrease with annealing. It is also observed that the nitrogenated film is more sensitive to annealing than the non-nitrogenated film, in agreement with earlier work by Burden et al. [11]. This is reinforced by the electrical data shown in Fig. 4, which shows very little change in the non-nitrogenated film at the 250°C, 5 min anneal. However the nitrogenated film annealed at 250°C shows an initial decrease followed by an increase in conductivity at higher temperature levels. The variation at 250°C could be explained if the conduction mechanism is defect-related, for example hopping conduction. The reordering of the lattice and the ensuing reduction in the density of electronically active defects would lower the film conductivity.

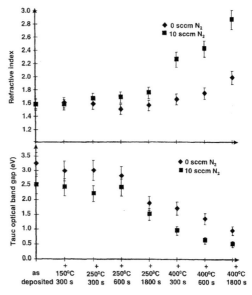

Fig. 3. Variation of refractive index and Tauc optical band gap with annealing conditions.

(a)

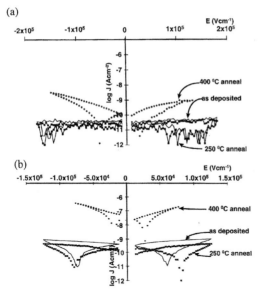

(b)

Fig. 4. J versus E characteristics for films grown on the earthed electrode and annealed to different temperatures: a) 0 sccm N_2, b) 10 sccm N_2.

Hysteresis in the J/E characteristics is also apparent. This is most likely due to trapping/detrapping of carriers with applied electric field.

At longer anneal durations at 400°C, large increases in conductivity are observed. This can be attributed to the onset of graphitisation of the film, because isolated islands of sp^2 are being formed and connected. This would also result in a corresponding decrease in the band gap as π states are introduced.

4. Conclusions

In this study both polymer-like and diamond-like hydrogenated carbon films have been grown using both electrodes of a rf-PECVD system. Both materials have been optically and electrically characterised. It has been observed that the diamond-like films follow Poole-Frenkel conduction which appears to hold even when nitrogen is added to the deposition chamber in order to form a-C:H:N. The variations in conductivity, optical band gap and refractive index may be explained by several factors, including electronic defect passivation and sp^2 bond formation. Electronic doping may also be occurring at low

nitrogen flows. Future work using techniques such as activation energy calculations from conductivity versus temperature measurements will help in understanding the conduction mechanism better.

The polymer-like films grown on the earthed electrode do not exhibit a Poole-Frenkel characteristic, and appear to be most likely candidates for electronic devices. The addition of nitrogen appears to alloy the film, reducing the band gap but not affecting the refractive index, and causing an increase in the conductivity of the film.

The effect of annealing the films grown on the earthed electrode seems to be both to reduce the density of electronically active defects at moderate temperatures (250°C), and also to graphitise the film at higher temperatures (400°C). The nitrogenated films are more sensitive to annealing, so nitrogen may be aiding the formation of sp^2 states at higher temperatures. Non-nitrogenated films have been shown to be thermally stable to 250°C. Therefore, it is possible that these films may be of use as future passivation layers for microelectronic circuits, where low conductivity, low dielectric constants and good thermal stability are of paramount importance.

Acknowledgements

The authors wish to acknowledge the EPSRC for financial support (GR/L09202).

References

[1] Street RA. Amorphous Hydrogenated Silicon. Cambridge University Press, 1991, Chapter 2.
[2] Silva SRP, Amaratunga GAJ, Constantine CP. J Appl Phys 1992;72:1149.
[3] Silva SRP, Robertson J, Amaratunga GAJ, et al. J Appl Phys 1997;81:2626.
[4] Franceschini DF, Achete CA, Friere FL. Appl Phys Lett 1992;60:3229.
[5] Kaufmann H, Metin S, Saperstein DD. Phys Rev B 1989;39:13053.
[6] Koidl P, Wild C, Wagner J, Ramsteiner M. Mat Sci Forum 1989;52:41.
[7] Silva SRP, Khan RUA, Anguita JV, et al. Thin Solid Films (in press).
[8] Barklie RC, Collins M, Silva SRP. Diamond Relat Mater 1998;7:864.
[9] Penn DR. Phys Rev 1962;128:2093.
[10] Frenkel J. Phys Rev 1938;54:647.
[11] Burden AP, Mendoza E, Silva SRP, Amaratunga GAJ. Diamond Relat Mater 1998;7:495.

Pergamon

Carbon 37 (1999) 781–785

CARBON

Doping of diamond

R. Kalish*

Physics Department and Solid State Institute, Technion-Israel Institute of Technology, Haifa 32000, Israel

Received 16 June 1998; accepted 3 October 1998

Abstract

Diamond is a wide-bandgap semiconductor with unsurpassed physical and chemical properties. When doped, semiconducting diamond can lead to the realization of electronic and optoelectronic devices with exceptional properties. Diamond can now be doped p-type, with boron, both during CVD diamond film growth and by ion-implantation, and n-type with phosphorus during CVD growth. This paper reviews the current status of diamond doping and describes the electronic properties of the doped layers. Some potential applications of doped semiconducting diamond are described. © 1999 Elsevier Science Ltd All rights reserved.

Keywords: A. Diamond; B. Doping; CVD; Implantation; D. Electronic properties

1. Introduction

The covalent bonding of carbon atoms in diamond (sp^3 bonds) are extremely strong and short, thus giving diamond unique physical, chemical and mechanical properties unmatchable by any other material [1,2]. These include: record high thermal conductivity, extremely high electric break down field, very high electrical resistivity when undoped, yet electrically semi-conducting at practically any desired value when doped with boron (p-type), high carrier mobilities and a negative electron affinity (NEA).

It is thus obvious that diamond, having such outstanding properties, is the material of choice for many applications. Most of these have, however, emerged only recently due to the discovery that diamond can be deposited in the laboratory by various CVD technologies. When dopant atoms are introduced into the plasma during CVD growth, diamond films which contain the dopants can be grown. Indeed p-type diamond with good control on electrical conductivity has been obtained by this method. Most recently unambiguous n-type diamond has been obtained by the introduction of PH_3 into the growth chamber. The electrical properties of the so obtained n-type layer are as yet rather poor; nevertheless the finding that P does induce a donor level located 0.46 eV below the conduction band is of great importance as it paves the way to obtaining also

n-type diamond. p- and n-type doped diamond is required for the realization of diamond-based bipolar electronic devices and is most desirable for the realization of electron emitting devices.

Diamond can also be doped by ion-implantation; however, ion-implantation in diamond is complicated by the metastability of the sp^3 diamond bonding with respect to the stable sp^2 bonding configuration of graphite. Hence damaged diamond, which always accompanies ion-implantation, may convert upon annealing to graphite.

The wide band-gap of diamond makes the realization of ohmic contacts to diamond difficult. The high resistivities, often encountered in lightly or moderately doped diamond, make meaningful electrical measurements, such as Hall effect measurements, very difficult. Nevertheless, good control on p-type diamond has been achieved both by CVD growth [3] and by ion-implantation [4]. Most recently, unambiguous n-type conductivities in phosphorus doped homoepitaxially grown diamond were reported [5,6].

2. Semiconducting doped diamond

2.1. p-type diamond

Boron doped p-type diamond exists in nature. The growth of boron doped diamond films by CVD techniques has been achieved by adding B containing molecules to the

E-mail address: kalish@ssrc.technion.ac.il (R. Kalish)

0008-6223/99/$ – see front matter © 1999 Elsevier Science Ltd All rights reserved.
PII: S0008-6223(98)00270-X

gas mixture in either a microwave (MW) or in a hot filament (HF) reactor resulting in the growth of B containing p-type diamond films. As this method is now well established, it will not be further discussed here.

Doping of diamond by boron ion-implantation has also been accomplished by employing different implantation/annealing schemes. These were summarized by Kalish [7].

The Technion group [8] has shown that by performing consecutive implantation and annealing steps, all to doses below the critical dose for graphitization, excellent control on doping level (and hence on activation energy and hole mobility) can be achieved as shown in Fig. 1.

Best results as for Hall mobility, were obtained most recently [9] by annealing at high temperature (1450°C, 10 min) a high energy deeply B implanted type IIa diamond. This implantation and annealing has resulted in a 'delta doped' p-type layer located about 1.5 μm below the surface with the very high hole mobility of over 600 cm^2/V/s.

A different annealing scheme has been devised [10] to create heavily boron doped near-surface layers of diamond, as required for the realization of ohmic contacts to p-type diamond. The method consists of high dose low energy B implantations followed by graphitization and chemical etching. This procedure leaves, after the graphite removal, only the tail of the B distribution hence creating a shallow (~10 nm thick) heavily doped p-type layer on the sample surface. The p$^+$ doping of diamond by this high dose

implantation/annealing/etching technique is now widely used to realize excellent ohmic contacts to p-type diamond.

The successful doping of polycrystalline CVD diamond films by B ion-implantation [4,8,10] which have yielded results comparable to those obtained for single crystal type IIa diamonds (although with poorer electrical properties) indicates that the technology developed for implantation doping diamond single crystals is applicable also to implantation doping of CVD diamond films. This finding has, obviously, very important technological implications.

2.2. n-type diamond

Whereas p-type doping of diamond has been achieved by a variety of techniques, effective n-type doping remained inconclusive. Only most recently clear donor activity in phosphorus doped diamond has been demonstrated [5]. In contrast to the case of p-type doping, where Nature has provided information on the preferred dopant (B) to be used, and on the electrical properties which one may expect of the doped diamond, the situation regarding n-type doping is much more complicated. There was, up till recently, no clear evidence as to which potential donor atom will yield a shallow enough energy level in the gap to be sufficiently ionized at room temperature and there was no information how this atom can be introduced into an electrically active site in an undamaged diamond lattice. The question which ohmic contact to use for a n-type

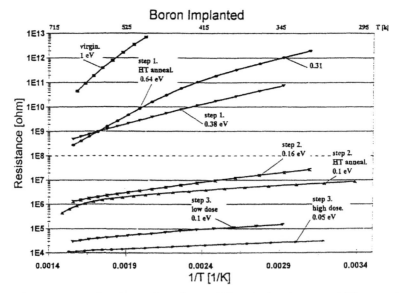

Fig. 1. Resistance vs. inverse temperature for B implanted CVD diamond. Different implantation doses and different annealing temperatures were employed. At low doping levels the well known activation energy for the B acceptor level in diamond is obtained. For higher B concentrations, 10^{20} and 2×10^{20} B/cm^3 achieved by consecutive implantation and annealing sequences; lower activation energies are obtained, signifying the transition to conduction in an acceptor band.

diamond is also still an open issue although evaporated Ti/Au contacts, which are nicely ohmic for p-type diamond, seem to work well also for n-type (P doped) diamond.

Several impurities are candidates to act as donors in diamond when occupying the proper lattice site: Group I elements (Li, Na) residing on interstitial sites [11] or group V elements (N, P, As) on substitutional sites were predicted to act as donors in diamond [12,13]. However, attempts to dope diamond with these using a variety of techniques did not yield convincing electron related conductivities. The conduction type has, in most experiments aimed at obtaining n-type diamond, been deduced from thermo-power (hot point) measurements. These, however, are only qualitative and are known to be unreliable when applied to highly resistive materials such as doped diamond usually is. 'Activation energies' have often been deduced from resistivity measurements over a rather narrow temperature range.

Thus, up till most recently no reports on n-type diamond which directly prove donor related electrical conductivity in diamond by Hall effect measurements have appeared in the literature. The breakthrough in obtaining n-type diamond came with the finding of the NIRIM group [6] who have grown CVD diamond which contains phosphorus dopants, and the detailed electrical characterization of the so grown diamond [5] which has conclusively demonstrated n-type conductivity. Details of the optimized growth conditions can be found in Ref. [6]. Secondary Ion Mass Spectroscopy (SIMS) was used to assess the actual P content in the films as well as the presence of other impurities (H, N etc.). It turned out that it is most important to grow the doped layer under such conditions which will minimize the H content in the film.

The dependence of the resistivity (ρ) on measurement temperature (T) for several P doped samples, differing in nominal P concentrations in the gas mixture during growth was measured over a wide temperature range (100–900 K). The results are plotted in Fig. 2 as log(ρ) vs. $1/T$. The fact that quite similar slopes are obtained for all samples proves that in this temperature range the conduction mechanism is thermally activated, with an activation energy of 0.46 eV, rather independent of growth conditions.

Hall effect measurements, with a magnetic field of ±0.7 T, were possible only for temperatures higher then about 230 K and even then, not for all samples, presumably due to poor carrier mobilities caused by the presence of defects in the doped layers. However, for the numerous samples that could be measured the temperature dependence of the Hall coefficient (R_h) was always *negative*, proving *n-type conductivity*. The carrier concentration $n(T)$ and the hole mobilities, as deduced from the Hall effect measurements for the 600 ppm sample, are plotted in Fig. 3 as log(n) or μ vs. $1/T$. When the carrier concentration data are fitted to the proper expression an activation energy of 0.46 eV is obtained. The electron concentration obtained for the 600

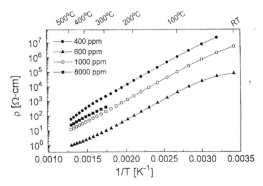

Fig. 2. Temperature dependence of the resistivity of n-type diamond, doped with different amounts (ppm) of phosphorus (300<800 K).

ppm sample at the maximal measurement temperature (600°C) is 3×10^{17} cm^{-3}. By comparing this value, which is a minimal value for the donor concentration, with the actual P concentration in this sample, as determined by SIMS (2.8×10^{18} P/cm^3), n-type doping efficiency of at least 10% is obtained for this sample.

The electron mobility data $\mu(T)$ for the 600 ppm sample in the temperature range that allows Hall measurements are also shown in Fig. 3. The mobility for this, as well as for all other samples, is always rather low (≤ 30 cm^2/V/s) and seems to be limited by the presence of defects.

Attempts to achieve n-type diamond by P ion-implantation have also been made [8,14]. Various contacts and various implantation/annealing sequences were tried out, yielding conductivities which were attributed in Ref. [14] to the presence of P in a donor state based on the 'hot-probe' test. Careful experiments on P ion-implanted type IIa diamonds employing a wide variety of implantation and annealing modes, including the methods which have yielded good p-type diamond by B implantations and

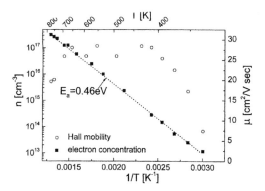

Fig. 3. Temperature dependence of the electron concentration and mobility as deduced from the Hall effect measurements.

those described in Ref. [14], were carried out by the Technion group [8]. These experiments, which were always accompanied by control experiments in which suitable inert ions were similarly implanted (to eliminate effects due to damage alone), did not yield electrical conductivities which could unequivocally be attributed to the presence of a donor state due to electrically active P.

An interesting observation found by the Technion group [8] is that, in contrast to the case of B implanted samples, for which annealing at the highest available temperatures (1450°C) has greatly improved the electrical properties of the sample, this does not seem to be the case for P doped diamond. High temperature annealing was found to wash out the little P related conductivity in P implanted samples. A similar behaviour was also found for homoepitaxial P doped CVD samples, for which excessive heating has drastically reduced the electrical conductivity.

Attempts to achieve n-type conduction due to the presence of group I impurities (Li and Na) in diamond have also been made [15]. However, no conclusive proof that these interstitial dopants yield useful n-type conductivities in diamond has been provided so far.

3. Potential applications of doped diamond

The application of semiconducting diamond has been reviewed [16]. It can be divided into two groups: One that requires the diamond to be relatively highly conductive, yet with no particularly stringent need for transport properties other then the low resistivity of doped diamond, while the second fully utilizes the semiconducting properties of diamond.

The first group of applications just takes advantage of the combination of the unique physical or chemical properties of diamond (or CVD diamond films) with the fact that doped diamond can conduct electrically. As is obvious from the above, the technology how to achieve such diamonds is well under control, hence most applications which have been practically implemented to date make use of heavily boron doped CVD diamond. Based on these, some temperature and pressure sensors as well as some Schottky devices are already commercially available [17].

A most promising application of conductive diamond is in the field of electrochemistry. Here there is often need for an electrode which can operate inertly and without deterioration in harsh chemical environments. Boron doped diamond has turned out to exhibit unsurpassed properties for such applications [18]. In Fig. 4 the I–V curves obtained with a B doped CVD diamond electrode in various (KI, KBr, and HCl) solutions are shown. The behaviour of the doped diamond electrode is much superior to that of the commonly used noble metal electrodes (Pt for example, which has until now been considered to be one of the best for such applications).

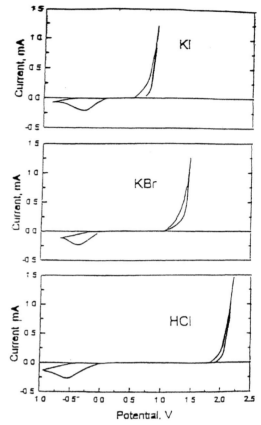

Fig. 4. C.V. of a highly B doped CVD diamond electrode in (a) 1 M KI; (b) 1 M KBr and (c) 1 M HCl. Scan rate 150 mV/s.

The great promise that diamond bears as a material for the fabrication of cold cathode or other electron emitting devices requires the diamond to be electrically conductive, with no need for an accurately known doping level. Here, however, the position of the Fermi level is of importance, hence n-type conductivities may be advantageous over the now used p-type diamond layers.

The second family of applications of doped diamond is in the field of electronic devices. Here the requirements on the material properties are much more demanding and no practical device has been realized as yet.

The unique electron emission properties of diamond are among the most promising applications of semiconducting diamond. Even though tremendous research efforts are devoted to this topic, no clear understanding of the physics which determined the electron emission from diamond emerges. It is however clear that for many applications, such as field emission (FE) from diamond surfaces, the injection of carriers from the backside of the device is essential as is the transport of electrons to the emitting surface, hence the diamond must be conductive. Most FE results which have been reported so far relate to p-type, B doped diamond. However, the advantages of using n-type diamond for this application is obvious.

4. Conclusion

In summary, the semiconducting properties of diamond can now be well controlled and start to find practical applications. While the situation regarding doping diamond p-type is satisfactory, and the technology for controlling the doping levels and the carrier transport properties is well at hand for both doping during CVD diamond growth and by ion-implantation, the situation regarding n-type conductivity in diamond is still at its infancy. The fact that P has been proven to have a donor level in diamond is encouraging and opens up many new research directions. Nevertheless, the donor level due to the P dopant is rather deep (0.46 eV) and may thus not be useful for some applications; hence the search for a shallower donor in diamond has to be continued and intensified. Nevertheless, thanks to its outstanding physical, electrical and chemical properties, semi-conducting diamond will, no doubt, play an important role in many novel emerging technologies.

Acknowledgements

Part of this work has been supported by a grant from the Israeli Ministry of Science.

References

[1] Field JE. The properties of natural and synthetic diamond. New York: Academic Press, 1992.

[2] Davis G, editor. Properties of diamond. INSPEC: The Institute of Electrical Engineering, 1994, and references therein.

[3] Glass JT, Dreifus DL, Fauber RE, Fox BA, Hartsell ML, Henard RB, Holms JS, Malta D, Piano LS, Tessmer AJ, Tessmer GJ, Wynands HA. Adv. New Diam. Sci and Tech. Tokyo: MYU, 1994:355.

[4] Fontaine F, Uzan-Saguy C, Kalish R. Appl Phys Lett 1996;68:2264.

[5] Koizumi S, Kamo M, Sato Y, Mita M, Sawabe A, Reznik A, Uzan-Saguy S, Kalish R. Diamond Rel Mater 1998;7:540.

[6] Koizumi S, Ozaki H, Kamo M, Sato Y, Inuzuka T. Appl Phys Lett 1997;71:1065–7.

[7] Kalish R. Ion-implantation in diamond: damage, annealing and doping. In: Paoletti A, Tucciarone A, editors. Proceedings of the International school of Physics 'Enrico Fermi' on 'Physics of Diamond'. IOS press, 1997:373.

[8] Ran B. Technion M.Sc. thesis, 1997, unpublished.

[9] Uzan-Saguy C, Kalish R, Walker RJ, Prawer S. Diamond Films & Tech 1998; in press.

[10] Kalish R, Uzan-Saguy C, Samoiloff A, Locher R, Koidl P. Appl Phys Lett 1994;64:2532.

[11] Kajihara SA, Antonelli A, Bernholc J, Car R. Phys Rev Lett 1991;66:2010–3.

[12] Zvanut ME, Carlos WE, Freitas Jr. JA, Jamison KD, Hellmer RP. Appl Phys Lett 1994;65:2287–9.

[13] Prawer S, Jamieson DN, Walker RJ Lee KK, Watt F, Kalish R. Diamond Films Technol 1996;6:351–58.

[14] Prins JF. Diamond Rel Mater 1995;4:580–5.

[15] Prawer S, Uzan-Saguy C, Braunstein G, Kalish R. Appl Phys Lett 1993;63:2502–4.

[16] Pan LS, Kania DR, editors. Diamond: electronic properties and applications. Boston, Dordreche, London: Kluwer Academic Publishers, 1995.

[17] Dreifus DL. In: Pan LS, Kania DR, editors. Diamond: electronic properties and applications. Boston, Dordreche, London: Kluwer Academic Publishers, 1995:371.

[18] Vinokur N, Miller B, Avyigal Y, Kalish R. J Electrochem Soc 1996;143:L238.

Carbon 37 (1999) 787–791

δ-Doping in diamond

M. Kunze[a,*], A. Vescan[a], G. Dollinger[b], A. Bergmaier[b], E. Kohn[a]

[a]*Department of Electron Devices and Circuits, University of Ulm, D-89069 Ulm, Germany*
[b]*Physics-Department E 12, Technical University of Munich, D-85747 Garching, Germany*

Received 17 June 1998; accepted 3 October 1998

Abstract

δ-Boron-doped homoepitaxial diamond films grown by microwave CVD were optimized for field effect transistor application to obtain steep profiles. The critical growth steps of the δ-doped device structures were analyzed and improved using mass spectrometry gas analysis, determining growth- and etch rates, hall-effect-measurements, elastic recoil detection and conductivity measurements. Optimized growth procedures were obtained and residual doping in the gate control layer was compensated using nitrogen. This results in a novel lossy dielectric Junction FET channel with high sheet charge activation and high drain current densities at moderate operation temperatures of 200°C. © 1999 Elsevier Science Ltd. All rights reserved.

Keywords: A. Diamond; B. Chemical vapor deposition; Doping; C. Mass spectrometry; D. Electronic properties

1. Introduction

The characteristics of diamond field effect transistors have been limited in the past by a reduced free carrier density, due to the high activation energy of the boron acceptor (370 meV at low doping densities). To solve this problem δ- or pulse-doped layer structures have been proposed [1] and first promising results have already been reported [2,3]. At high doping levels ($>10^{20}$ cm^{-3}) the

activation energy is reduced leading to almost 100% carrier activation. In a δ-doped structure the free carrier distribution is also expected to be considerably spread out beyond the doped layer, resulting in reduced ionized impurity scattering and high mobilities [1]. Therefore a δ-boron-doped channel field effect transistor with sheet concentration in the 10^{13} cm^{-2} range is expected to yield drain current densities even above 1 A/mm [3]. A schematic cross section of a δ-doped FET is shown in Fig. 1.

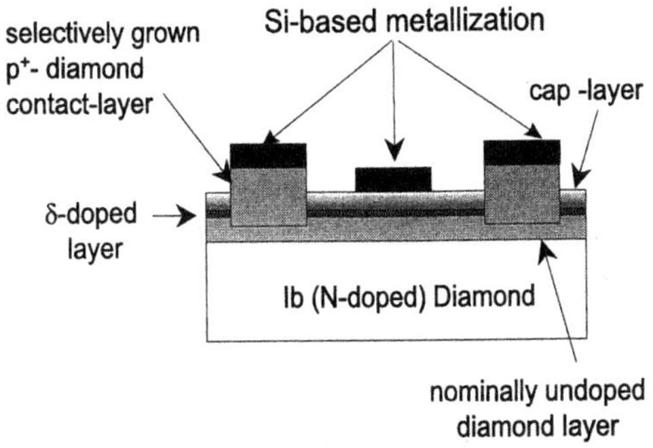

2. Experimental

The investigated structures were grown in a commercial diamond microwave-CVD reactor (ASTEX™, Fig. 2). The substrate material was highly nitrogen (*n*-type) doped single crystal synthetic diamond. Definition of mesa-structures was performed by selective epitaxy, using SiO$_2$ masks. The growth was performed in using 1.5% CH$_4$ in H$_2$ at a total pressure of 30 torr. First a 500-nm thick, nominally undoped buffer layer was grown at a substrate temperature of ~650°C. The growth rate was about 5 nm/min. For *p*-type doping a boron rod was inserted into the plasma for a short time (≥3s). Finally a 100 nm thick cap-layer was grown at a higher temperature of 750°C (11 nm/min) in order to minimize memory effects from the residual boron in the growth chamber. The ohmic contacts

selectively grown
p⁺- diamond
contact-layer

Si-based metallization

cap -layer

δ-doped
layer

Ib (N-doped) Diamond

nominally undoped
diamond layer

Fig. 1. Schematic cross section of a δ-doped-channel FET. The ohmic contact regions are defined by selective epitaxy of *p*⁺-doped diamond [4].

*Corresponding author.

0008-6223/99/$ – see front matter © 1999 Elsevier Science Ltd. All rights reserved.
PII: S0008-6223(98)00272-3

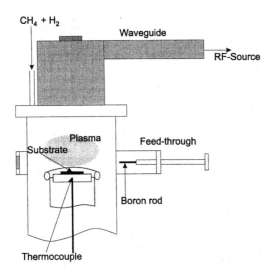

Fig. 2. CVD-System with solid boron rod as doping source. Typical growth conditions are 1.5% CH$_4$ in H$_2$ at a total pressure of 30 Torr, substrate temperature of approx. 650°C and 700 W RF-Power at 2.45 Ghz.

for this pulse doped structure were fabricated using selectively grown p^+-doped contact regions, as previously reported [4]. Both ohmic and Schottky contact metallization were sputter-deposited in a commercial ion beam reactor using highly temperature stable Si/WSi$_x$N$_y$/Ti/Au metals. The main growth conditions for the presented device structure are as follows.

1. The design with highly activated δ-channel imposes tight limits of the width of the δ-doped channel. For a sheet concentration of 10^{13} cm^{-2} a width of the δ-doped layer in the range of 1 nm is needed to achieve full activation. For both narrow doping profile and reproducible doping time (≥3 s to minimize error) small growth rates are required during the doping step. Analyzing the growth rates (Fig. 3), growth temperatures of T_S=600°C at 1.75% CH$_4$ in H$_2$ will be needed or reduced CH$_4$-flux (for example 0.5% CH$_4$ in H$_2$ leading to a reduced growth rate of $r≈95$ nm/h). Therefore, the growth procedure of the doped layer determines the growth parameters of the nominally undoped buffer layer avoiding a growth interrupt. The activation energy of E_a=14.8 kcal/mol (Fig. 3) indicates that dissociation of hydrogen from the diamond surface limits the growth rate [5].

2. A further critical feature of the structure lies in the cap layer. Here extremely low doping is needed to ensure a low leakage Schottky-gate contact. Also residual doping in the cap layer will inhibit modulation of the channel charge. Therefore after growing the doped layer a

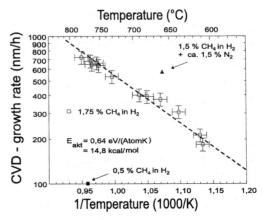

Fig. 3. Growth rates of the CVD-System.

growth interrupt was performed to remove residual boron out of the chamber.

The main influence on achieved activation energy of the devices can be seen in Fig. 4. Changing the doping time from 30 s down to 3 s leads to an increased activation energy from 0 meV up to 260 meV.

Optimization of the gate control layer requires a lot of research into growth conditions: usually each growth-start begins with some seconds or a few minutes of pure hydrogen plasma to ensure the required not-reconstructed, hydrogen saturated diamond surface [6]. The diamond-etching effect of such procedures are well known. At the growth-interrupt and at the restart after growth of the narrow pulse doped layer the etch rates and also gas flows have to be known to avoid any negative effect on the pulse

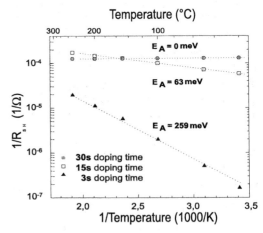

Fig. 4. Activation energies of pulse doped diamond films for different doping times.

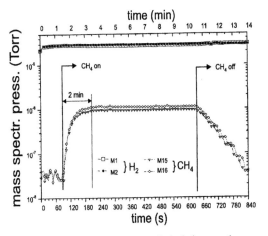

Fig. 5. Mass spectroscopy gas analysis during growth.

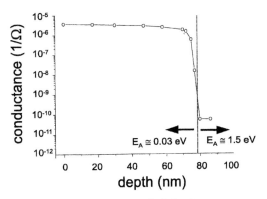

Fig. 7. Conductivity profile of a δ-doped structure.

doped layer. Fig. 5 shows the mass spectroscopy analysis of CH_4 and H_2 during growth. After 3 min of stop of the CH_4-flux the concentration falls off at the background value and therefore further diamond growth occurs with starting of the etch process of diamond due to increasing hydrogen excess. The analysis of the etch rates of diamond in pure hydrogen (Fig. 6) reveals no significant effect on the pulse doped layer for the used 30 s 'outgrowth-time' and 30 s pretreatment in pure hydrogen before restart. Also variations on outgrowth time revealed no significant influence on activation energy of the devices.

3. Results

The conductivity profile of such a film is shown in Fig. 7 and was obtained by sequentially etching the top layer and measuring the conductivity after each step. A deep

decrease of conductivity over several orders of magnitude is observed over a depth range less than 15 nm, indicating that the conductivity is dominated by the doping pulse. This is confirmed by the activation energy, which remains basically constant at about 30 meV during the profiling. After etching the pulse doped layer, an activation energy of 1.3 eV for the undoped diamond buffer layer was extracted. Elastic recoil detection measurements have confirmed the steep doping profile, with a FWHM less than 10 nm. Elastic recoil detection on a pulse doped diamond film reveals similar results on the incorporated boron profile (Fig. 8). A full width at half maximum of only less than 6 nm was achieved. Nevertheless, not negligible residual doping in the cap layer leads to a sheet carrier concentration of $n_s \approx 1.6 \times 10^{13}$ cm^{-2}. The high sheet concentration towards the surface is already too high to modulate the peak carrier concentration and furthermore the high surface concentration limits the rectifying Schottky contacts. Similar results can be extracted from

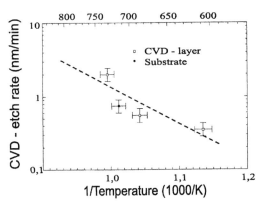

Fig. 6. Etch rates of diamond in pure hydrogen plasma.

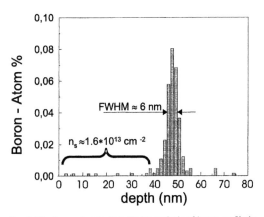

Fig. 8. Elastic recoil detection (ERD) analysis of boron profile in pulse doped diamond film.

the conductivity profile (Fig. 7) where the slow decrease before etching the δ-doped layer itself is due to the background doping in the cap layer.

Using the extracted doping profile from ERD-measurements and the conductivity-profiling, leading to a model with two conductivity paths, Hall measurements on such a pulse doped film (Fig. 9) can be approached. The first path with high carrier concentration (1.9×10^{21} cm^{-3}, assuming a pulse doping width of 10 nm), therefore low activation energy of $E_a \approx 3$ meV and low mobility of $\mu \approx 3$ cm^2/Vs dominates the range below $T = 170$ K. These values are in good agreement with reported values extracted from highly doped samples [7], indicating hopping-conductivity. At the temperature range above $T = 170$ K an activation energy of $E_a \approx 180$ meV can be extracted. Using an active layer thickness of 100 nm, a film-thickness of 600 nm and from ERD-measurements determined nitrogen background-concentration of about 10^{18} cm^{-3} (therefore compensating factor $K \approx 0.25$), a doping concentration of $N_A \approx 10^{19}$ cm^{-3} results. Therefore no change in dominant conduction mechanism may be assumed, because of the different doping concentrations and activation energies. The small activation energy at $T < 170$ K may be due to the highly doped δ-layer and the high activation energy at $T > 170$ K with reduced carrier concentration due to the conductivity of the cap layer with parasitic doping concentration towards the surface.

From these investigations the properties of our δ-doped layers may be summarized as follows: (i) the peak doping concentration is above 10^{19} cm^{-3} or even higher leading to the desired reduction of the activation energy and dominating the conductivity of the whole structure. (ii) The residual doping in the cap layer contributes only negligibly to the overall conductivity, however it inhibits the formation of a good Schottky-gate and prevents

modulation of the channel charge. This was actually confirmed in FET structures which showed only negligible current modulation using a Schottky-gate on the boron-contaminated cap-layer.

We propose a FET structure, where the doping tails towards the surface are cut by using the only choice of n-type doping nitrogen with an activation energy of 1.7 eV. In the presence of boron it compensates p-type doping and may therefore be used to adjust the doping profile and the effective doping concentration in the cap layer. Depending on the amount of built-in nitrogen concentration overcompensation may occur. This will lead to a transition region from p-δ-channel to n-type on top and therefore lead to a lossy dielectric region in between. Fig. 10 demonstrates the effect of nitrogen incorporation on the activation energy of the device changing only the purity of the used hydrogen. ERD-measurements revealed for the hydrogen with purity 5.3, a nitrogen concentration of about 16 ppm and for purity 6.0 a decreased concentration of about 2 ppm corresponding to dramatically reduced activation energy. Borst et al. have demonstrated in the past [7], that a structure consisting of a boron doped layer grown on n-type nitrogen doped single crystal substrates yield the electrical behaviour of a p/n-junction. It should therefore be possible to use such a diode in a Junction-FET instead of a Schottky-gate.

In a processed device structure a 100-nm thick nitrogen doped layer was placed on top of the δ-doped layer. This was done by selective epitaxy and by adding nitrogen to the process gas during the growth of the cap layer. The resulting output characteristics of this FET structure is shown in Fig. 11 for a gate length of 20 μm. Due to the

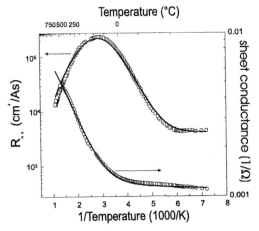

Fig. 9. Hall effect measurement and fit curves for pulse doped diamond film.

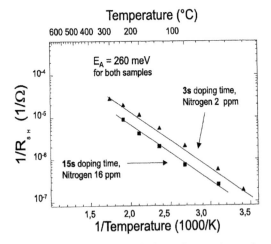

Fig. 10. Analysis of the effect of nitrogen incorporation on the activation energy of pulse doped devices using different hydrogen purity. (ERD-data by A. Bergmaier, Techn. University of Munich).

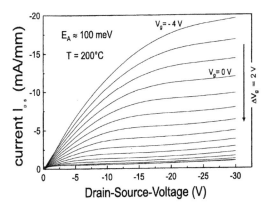

Fig. 11. Output characteristics of a Junction FET with nitrogen doped cap layer, with a gate length of 20 μm.

short doping time of only 3 s, the activation energy is slightly increased to 140 meV. Nevertheless a maximum drain current density of 20 mA/mm can be achieved already at 200°C. The device can be completely turned off, indicating that in this structure the complete channel charge in the δ-peak can be depleted. At 10 μm gate length the maximum drain current reaches even 40 mA/mm, however pinch off could not be achieved. The maximum extrinsic transconductance is 1.4 mS/mm. To our knowledge these are the highest values reported so far for boron doped diamond field effect devices at this gate length and such low temperatures. Scaling these results down to a gate length of 1 μm we expect that the current densities may be increased above 200 mA/mm and therefore comparable to other wide band-gap semiconductors.

In conclusion these preliminary results demonstrate clearly that using an *n*-dopant like nitrogen, a Junction FET may be realized on diamond films. The undesired residual doping in the gate control layer can be reduced and problems related to the formation of a Schottky-gate eliminated. Using this concept together with a δ-boron-doped channel the activation energy can be considerably reduced, leading to high current densities. The performance may be further increased by reducing the gate resistance using more shallow *n*-type doping. Therefore phosphorous may be an alternative. Recently published results on phosphorous as a donor in diamond [8], indicate a reduced activation energy around 0.5 eV.

Acknowledgements

This work was supported by the Deutsche Forschungsgemeinschaft, carried out under the frame of the trinational 'D-A-CH' consortium.

References

[1] Anda Y, Ariki T, Kobayashi T. Quantum analysis of hole distribution in multiple-delta-doped diamond with a deep level impurity level. Jpn J Appl Phys 1996;35:3987–90.

[2] Shiomi H, Nishibayashi Y, Toda N, Shikata S. Pulse-doped diamond *p*-channel metal semiconductor field-effect-transistor. IEEE Electr Dev Lett 1995;16(1):36–8.

[3] Vescan A, Gluche P, Ebert W, Kohn E. High-temperature, high-voltage operation of pulse-doped diamond. MESFET. IEEE Electr Dev Lett 1997;18(5):222–4.

[4] Vescan A, Gluche P, Ebert W, Kohn E. Selectively grown ohmic contacts to δ-doped diamond films. Electr Lett 1996;32(15):1419–20.

[5] Nishimori T, Sakamoto H, Takakuwa Y, Kono S. Methane adsorption and hydrogen isothermal desorption kinetics on a C(001)-(1×1) surface. J Vac Sci Technol A 1995;13:6.

[6] Anthony TR. Metastable synthesis of diamond. Vacuum 1990;41:1356–9.

[7] Borst TH, Strobel S, Weis O. High-temperature diamond *p–n* junction: B-doped homoepitaxial layer on N-doped substrate. Appl Phys Lett 1995;76(18):2651–3.

[8] Kolzumi S, Kamo M, Sato Y, Mita A, Sawabe A, Reznik A, Uzan-Saguy C, Ran B, Kalish R. Phosphorous doped *n*-type diamond. Diamond Related Mater 1998;7:540.

Pergamon

Carbon 37 (1999) 793–799

CARBON

Electronic properties of single crystalline diamond surfaces

L. Ley*, R. Graupner, J.B. Cui, J. Ristein

Institut für Technische Physik, Universität Erlangen, Erwin-Rommelstr. 1, 91058 Erlangen, Germany

Abstract

Key parameters of the surface electronic structure of diamond (111), such as surface core level shifts, surface state dispersion, band bending, and electron affinity, are presented and discussed. The first actual value for the negative electron affinity of diamond (111):H 1×1 is measured as $\chi = -1.27$ eV. A model is given that describes the variation in χ with hydrogen coverage quantitatively in terms of C–H bond dipoles. © 1999 Published by Elsevier Science Ltd. All rights reserved.

Keywords: A. Diamond; D. Electronic properties

1. Introduction

Thin-film diamond is readily deposited by plasma CVD methods on silicon substrates from CH_4 or other hydrocarbons highly diluted in H_2 [1]. Under proper nucleation conditions it is even possible to synthesize diamond films on Si(100) in which individual crystallites are in registry with the underlying silicon lattice in such a way that the term 'heteroepitaxy' is sometimes used [2].

Given this situation it is clear that first electronic devices based on CVD diamond have been realized [3]. A particularly interesting development in this respect is the so-called surface-MESFET, a field-effect transistor that relies on the switching of a highly conductive surface layer that is apparently present on a hydrogen saturated diamond surface [4,5]. The nature of this conductive surface layer is unknown as yet and it appears to be unique to H-covered diamond.

Hydrogen termination also plays a crucial role in another unique property of diamond, namely negative electron affinity (NEA). The electron affinity χ is the energy barrier an electron at the conduction band edge E_c has to overcome when it tries to leave a semiconductor into the vacuum: $\chi = E_{vac} - E_c$, where E_{vac} is the energy of the vacuum level. χ is usually of the order of 4 eV. However, in diamond χ can attain negative values making diamond a highly effective electron emitter provided

electrons are in the conduction band. Despite the fact that NEA was observed almost twenty years ago [6] a value for $\chi < 0$ has only recently been determined as will be discussed below [7]. From these brief examples it is obvious that the surface electronic properties of diamond are quite different from other semiconductors and deserve therefore particular attention. We shall give in this paper a brief overview of our recent work in this area which covers all aspects of the electronic and some structural properties of the diamond (111) surface.

2. Experimental

For our investigations we used slightly B-doped ($\sim 10^{16}$ cm^{-3}) single crystalline diamond with resistivities of about 1 Ωcm. Form the Boron concentration and the energy of the acceptor level (0.36 eV above the valence band maximum E_v) the Fermi level position in the bulk is calculated to lie 0.32 ± 0.01 eV above E_v. The samples with typical dimensions of 3×5 mm^2 were plasma polished so as to yield atomically flat, highly ordered, and hydrogen terminated surfaces as judged by sharp and low background LEED (low-energy electron diffraction) diffraction patterns and the dominance of dispersing features in the angle-resolved photoelectron spectra (ARPES) [8].

As experimental methods we employed a variety of photoelectron spectroscopies, some of them using synchrotron radiation from the BESSY storage ring in Berlin. The work function was measured using the Kelvin probe method and LEED was employed to monitor the

---

*Corresponding author. Tel.: +49-9131-857090; fax: +49-9131-857889.

E-mail address: lothar.ley@physik.uni-erlangen.de (L. Ley)

0008-6223/99/$ – see front matter © 1999 Published by Elsevier Science Ltd. All rights reserved.
PII: S0008-6223(98)00273-5

reconstruction of the surface. All experiments were carried out in ultrahigh vacuum.

Because one of the main objectives of our work was to study the influence of hydrogen on the surface structure annealing experiments up to temperatures of about 1200°C were used to drive the hydrogen off. This requires accurate temperature measurements which are notoriously difficult to perform on diamond. Thermocouples cannot be attached directly to the crystals and optical pyrometry does not work because diamond does not emit over most of the infrared spectral regime. We have therefore employed the frequency shift of the zone center optical phonon mode as a measure of temperature. When measured by Raman spectroscopy this frequency allows an absolute temperature determination with an accuracy of ± 10 K [10].

3. Results and discussion

3.1. C1s core levels

Figures 1–4 show a series of C1s core level spectra taken at the synchrotron with a resolution of about 150 meV [11]. The energy scale is given relative to the bulk component B, i.e. the emission of C1s electrons belonging to the bulk of diamond. The spectrum of Fig. 1 is that of a freshly plasma hydrogenated surface that has been carried from the plasma reactor through air to the spectrometer. The LEED pattern is that of an unreconstructed 1×1 surface. A second C1s component at 0.72 eV higher binding energy, S_A, is visible which is due to C atoms at the surface with a chemical environment that differs from

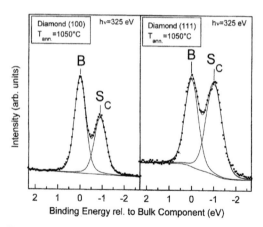

Fig. 2. C1s photoelectron spectra of the hydrogen-free, 2×1 reconstructed diamond (100) and (111) surfaces. The surface component S_C is due to atoms involved in π-bonds.

that of bulk diamond (surface core level shift). As the origin of S_A we identify polyhydride (C–H$_x$, $x=2,3$) configurations which are not very tightly bound to the surface because they desorb around 400°C. Whether we are dealing here with the trihydride terminated surface in which each dangling bond at the (111) surface is terminated by a C–H$_3$ group is questionable at best. Because the intensity of the surface component varies strongly with the plasma conditions (see Fig. 1) we are more inclined to attribute it to physisorbed carbohydride clusters on an otherwise monohydrogenated surface.

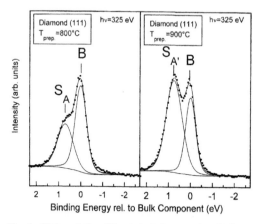

Fig. 1. High-resolution C1s photoelectron spectra of plasma hydrogenated single crystal diamond (111) surfaces. The photon energy is 325 eV which yields surface sensitive spectra with a sampling depth of a few Å. Note the large difference in the surface component S_A as a function of nominal sample temperature T_{prep} during plasma treatment.

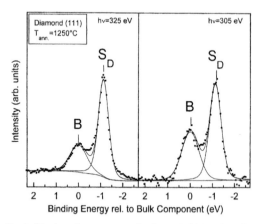

Fig. 3. C1s photoelectron spectra of a diamond (111) surface after partial graphitization. The photon energies were chosen such that the left spectrum ($h\nu = 325$ eV) is very surface sensitive and the right spectrum ($h\nu = 305$ eV) somewhat less so. The change in the intensity ratio of peaks S_D and B indicates that the graphitization takes place at the surface.

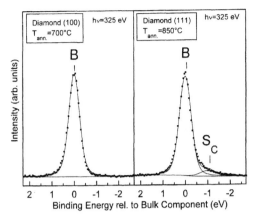

Fig. 4. C1s photoelectron spectra of monohydrogenated diamond (100) and (111) surfaces.

After annealing the surface above about 900°C the hydrogen desorbs and the surface reconstructs yielding a characteristic 2×1 LEED pattern. The reconstruction involves a massive rearrangement of the surface atoms such that next nearest neighbors become nearest neighbors and form chains of carbon atoms that are connected by double bonds ($\sigma + \pi$). This π-bonded chain reconstruction is similar to that observed on Si(111) after cleavage [12]. The atoms involved in the π-bonds give rise to the surface C1s component S_C in Fig. 2 at −1.0 eV which is therefore a clear signature of π-bonds. That this is indeed so is shown in Fig. 3 where a diamond surface was partially graphitized after prolonged annealing at 1250°C. The surface component S_D which corresponds to S_C is now the dominating line because in graphite all atoms are involved in π-bonds.

Between the polyhydride covered and the hydrogen-free reconstructed surface the surface is in a monohydrogenated phase in which each dangling bond is terminated by one H atom. The signature of this surface is the spectrum of Fig. 4 which shows except for a small indication of S_C only the 'bulk' C1s line B. This implies that any chemical shift due to the C–H bond is smaller than the linewidth of about 0.45 eV. Indeed, a detailed analysis of the evolution of the core level spectra during annealing yields a chemical shift of −0.15 eV for the C atoms bonded to H at the surface of diamond [11].

Virtually identical spectra with the same chemical shifts have been measured for the diamond (100) surface as demonstrated in Figs. 2 and 4.

3.2. Surface states on the (111) surface

A crucial question for any surface electronic structure of a semiconductor is that of surface states in the gap because they are likely to pin the Fermi energy E_F at the surface

away from its bulk position and cause therefore a surface band bending. They furthermore act as efficient recombination centers for carriers. Angle-resolved photoelectron spectroscopy (ARPES) is the method of choice to determine the energy and dispersion of surface states [8,9]. A comprehensive study of H-terminated (111) and (100) surfaces by ARPES has shown that these surfaces are free of occupied surface states in the fundamental gap [8,9]. However, there is evidence from C1s absorption spectra that empty surface states do indeed fall into the upper half of the fundamental gap of diamond [13].

Of particular interest in this respect is the H-free 2×1 reconstructed (111) surface. According to most calculations this surface is expected to be metallic, i.e. there should be at least one surface state that crosses E_F as indicated by the line in Fig. 5 [14]. The experimentally determined dispersion of the surface state (solid squares in Fig. 5) follows this calculations over most of the surface Brillouin zone but it does not cross E_F. In fact the point of closest approach at $\bar{K}$ is 0.45 eV and the important conclusion of this measurement is thus that the diamond (111) 2×1 surface is *semiconducting* with a gap of at least 0.45 eV. Whether the discrepancy between theory and experiment is due to computational shortcomings (improper treatment of electron–electron correlation) or a deficiency of the structural model on which the calculations were based remains to be seen.

3.3. The magnitude of NEA on diamond (111): H

The signature of NEA on the hydrogen terminated

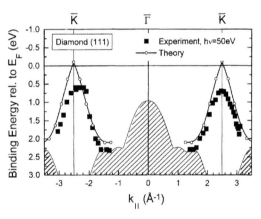

Fig. 5. Surface-state dispersion along the high symmetry directions $\bar{K}-\bar{\Gamma}-\bar{K}$ in the surface Brillouin zone of a hydrogen-free, 2×1 reconstructed diamond (111) surface. The open circles represent the calculation of Kern et al. [14] and the solid squares are the experimental results. The hatched regions represent the projected bulk bands. Note that the experimentally determined dispersion remains well below E_F whereas the calculation has the surface band cross the Fermi level.

diamond surfaces [both (111) and (100)] is a high-intensity tail of secondary electrons in a photoelectron spectrum that extends down to energies corresponding to the bottom of the conduction band E_c because electrons at E_c can still escape into vacuum. The reason why this intensity is so high has been explained in [15]. Clearly, such an observation gives no information about the magnitude of $\chi = E_{vac} - E_c < 0$, i.e. the energy difference between E_c and the vacuum level E_{vac} that lies below E_c. However, $\chi < 0$ can be measured by combining the work function $\Phi = E_{vac} - E_F$ with a determination of $E_F - E_v$ at the surface:

$$\chi = E_{vac} - E_c = E_{vac} - E_F + E_F - E_v - E_c + E_v$$
$$= \Phi + (E_F - E_v) - E_g \qquad (1)$$

where $E_g = 5.47$ eV is the gap of diamond.

The experiment designed to measure χ as a function of hydrogen coverage is performed as follows [7]. Starting with a fully hydrogenated surface, at first only changes in χ are measured as a function of annealing time by combining changes in Φ with the change in surface band bending $E_F - E_v$. The latter one is measured by the change in C1s binding energy relative to E_F which constitutes the natural reference energy for any photoelectron spectra. We thus have:

$$\Delta\chi = \Delta\Phi + \Delta(E_F - E_v) \qquad (2)$$

The result of these measurements for an isothermal annealing sequence at 1000 K is shown in Fig. 6. The open squares give the change in $E_F - E_v$ as derived from the C1s binding energy and the solid squares the changes in χ which equal $\Delta\Phi$ corrected by $\Delta(E_F - E_v)$.

In a second step the changes in χ are converted into absolute values by determining the point where $\chi = 0$. This point is marked by a cross in Fig. 6 and it was obtained by monitoring the photoelectron yield near the threshold of 5.5 eV as a function of annealing time. Fig. 7 shows three characteristic yield spectra in which the photoelectron count rate is plotted vs. photon energy. The signature of a NEA surface is a steeply rising yield that starts for photon energies that are equal to the gap (for details see [15]). When the surface turns to positive electron affinity (PEA) electrons at E_c encounter a barrier and are thus unable to leave the diamond; the yield remains low near E_g. Thus according to the spectra of Fig. 7 the switch from NEA to PEA occurs between 42 min and 54 min of annealing and that marks the point in Fig. 6 where $\chi = 0$ with an accuracy of ± 0.065 eV. With $\Phi|_{\chi=0} = E_g - (E_F - E_v)$ a value for $E_F - E_v|_{\chi=0}$ follows from the known value of $\Phi|_{\chi=0}$. In this way the values for χ and $E_F - E_v$ are placed on absolute scales as shown by the ordinates of Fig. 6. It is only by this rather involved method that values for $\chi < 0$ can be determined reliably and with high accuracy. The resulting surface band diagram is given in Fig. 8. The electron affinity changes by a total of 1.65 eV form -1.27 eV for the hydrogen saturated surface to $+0.38$ eV for the hydrogen-free, 2×1 reconstructed surface. At the same time does the surface barrier, i.e. $E_F - E_v$ at the surface, increase by 0.2 eV from 0.68 to 0.88 eV. The surface layer is thus in both cases depleted by carriers which are, of

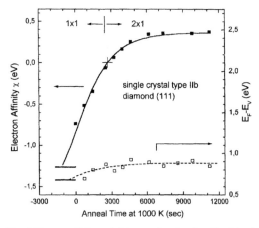

Fig. 6. Electron affinity χ (solid squares) and surface Fermi level position $E_F - E_v$ (open squares) as a function of annealing time of a diamond (111) surface kept at 1000 K. The solid line is a fit to the data assuming first-order desorption kinetics for H. The transition point from a 1×1 to a 2×1 surface is indicated.

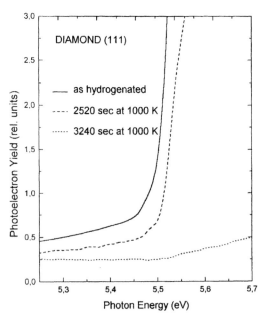

Fig. 7. Total photoelectron yield spectra taken at different times during the annealing sequence of Fig. 6.

Diamond (111) 2x1

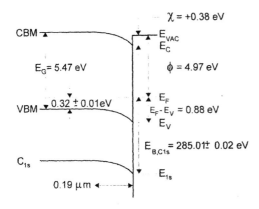

Diamond (111) 1x1:H

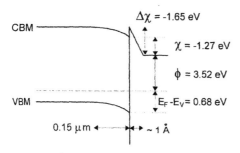

Fig. 8. Band diagram for the monohydrogenated unreconstructed (lower panel) and the hydrogen-free 2×1 reconstructed diamond (111) surface.

course, holes in the slightly B-doped (10^{16} cm^{-3}) sample with a bulk Fermi position that was calculated to lie 0.32 ± 0.01 eV above E_v.

An increase of downward band bending is generally observed after hydrogen desorption on (111); the actual values differ, however, considerably [6,17,18]. As will be shown in a forthcoming publication [19] this is due to the temperature used to drive off the hydrogen. Mild annealing at low temperatures as described here leads to a small additional band bending whereas shorter annealing at consequently higher temperature leads to $E_F - E_v = 1.42$ eV. If hydrogen is desorbed by electron bombardment, no increase in band bending is measured at all [20]. These observations imply that the change in band bending is not directly related to hydrogen desorption but rather due to charged defects that are created in parallel during thermal treatment. Indeed, the concentration of surface charge necessary to induce the observed band bending is only

about 10^{-3} of a monolayer and an incipient graphitization at a corresponding number of surface sites could easily account for it.

We argue that the lowering of χ is due to a dipole double layer set up by the polar $C^{\delta-}-H^{\delta+}$ bonds at the surface with C slightly more electronegative than H. Such a dipole layer causes a potential step ΔV perpendicular to the surface over a distance of the order of the C–H bond length of 1.1 Å. The potential step in turn implies that E_{vac} is lowered relative to E_c by $e\Delta V$ compared to its value without the dipole layer. Because ΔV depends on the areal density n and the magnitude p of the dipoles we have

$$\chi - \chi_{max} = -e\Delta V = -\frac{epn}{\varepsilon_0} g(n) \tag{3}$$

where ε_0 is the dielectric constant of free space. The function g which depends on n takes the interaction of dipoles into account, with the result that the contribution of each dipole to ΔV is reduced for high dipole densities. An expression for $g(n)$ with the polarizability α of the dipoles as a parameter can be obtained according to the calculation of Topping [21]:

$$g(n) = \left(1 + \frac{9\alpha n^{3/2}}{4\pi\varepsilon_0}\right)^{-1} \tag{4}$$

Considering first-order desorption kinetics of hydrogen from the diamond surface the dipole density as a function of isothermal annealing time t is given by

$$n(t) = n_0 e^{-A(T)t} \tag{5}$$

where n_0 is the areal density of H-atoms at the start of the annealing and equals thus the surface density of C-atoms of 1.81×10^{15} cm^{-2} for the (111) surface. Combining Eqs. (3)–(5) the change $\Delta\chi(t)$ can be calculated using $A(T)$ and p as free parameters. The best-fit result shown as a solid line in Fig. 6 is obtained for $A(1000$ K$)^{-1} = 1850$ s and $p = 1.45 \times 10^{-30}$ Asm (0.09 eÅ) using a polarizability α for the C–H dipoles of 1.28×10^{-40} Asm^2/V. This value was obtained from the refractive index of polyethylene which equals 1.5 using the Clausius–Mossotti relation [22]. It should be pointed out that a satisfactory fit of the experimental $\Delta\chi(t)$ values to Eq. (5) is not possible without taking the depolarization into account.

In the point charge approximation the dipole p is given by $p = d\Delta q$ where d is the C–H bond length and Δq the charge transfer from the more electronegative C to the H atom. Estimating Δq on the basis of Pauling electronegativities for C (2.55) and H (2.2) we obtain $\Delta q = 0.07e$ ([17,23]). Combining this value with the C–H bond length of 1.1 Å yields $p = 1.23 \times 10^{-30}$ Asm (0.08 eÅ). This value is in excellent if possibly somewhat fortuitous agreement with the dipole moment arrived at from our analysis of the electron affinity.

The electron affinities of different diamond surfaces were recently calculated [16]. The authors obtain $\chi =$

−2.03 eV for the hydrogenated, unreconstructed (111) surface and $\chi = +0.38$ eV after dehydrogenation and reconstruction.

In a second series of isochromal annealing experiments the sample was taken in steps through temperatures from 600 to 900°C. From a similar analysis of the electron affinity in terms of hydrogen coverage, the activation energy $E_H = 1.25$ eV and the prefactor $\nu_H = 1 \times 10^3$ s^{-1} for the rate constant $A(T) = \nu_H \times \exp(-E_H / kT)$ have been determined. The value for E_H is at the lower end of the activation energies reported for H-desorption from diamond which lie more around 3 eV. The reason for this discrepancy as pointed out by Schulberg et al. [24] lies in the fact that most authors use a fixed $\nu_H = 10^{13}$ s^{-1}. An even lower value of 0.8 eV was obtained by Nishimori et al. [25] for the desorption of H from diamond (100).

4. Summary and conclusions

We have characterized the diamond (111) surface in terms of C1s surface components, the dispersion of surface states, surface band bending, and electron affinity. The main results may be summarized as follows.

(i) There are four distinct surface phases that can be distinguished on the basis of the C1s core level spectra. The plasma hydrogenated surface exhibits an additional surface core level component S_A which is shifted by 0.72 eV to higher binding energy. We ascribe this component to atoms bonded in polyhydride (C–H$_x$, $x = 2.3$) configurations that desorb at temperatures of about 400–500°C. The monohydrogenated, 1×1 surface is characterized by an unresolved surface component with a shift of −0.15 eV. After hydrogen desorption and 2×1 reconstruction a third surface component S_C, with 1.0 eV lower binding energy is ascribed to C-atoms involved in the π-bonds of the 2×1 reconstruction. S_C evolves into S_D at −1.13 eV as the surface graphitizes for temperatures above 1200°C. (ii) The hydrogenated surface is free of occupied surface states in the gap. The hydrogen-free 2×1 reconstructed surface has a surface band in the gap that extends up to 0.45 below E_F. The surface is thus contrary to most calculations semiconducting with a surface gap of at least 0.45 eV. (iii) Both, the hydrogenated 1×1 and the hydrogen-free 2×1 surface exhibit hole depletion layers on our slightly B-doped samples. The surface band bending increases from 0.36 eV for the as prepared surface to 0.56 eV for the hydrogen-free 2×1 surface. This requires compensating positive surface charges of 1.25 and 1.73×10^{11} cm^{-2}, respectively, which cannot be accounted for by partially filled surface states. (The gap is too large in either case.) We therefore have to postulate positively charged defects of the order of 10^{-3} per surface atom to account for the band bending. It is furthermore evident from these measurements that the intrinsic diamond surfaces—hydrogenated or not—cannot account for the high surface conductivity that is the basis of the surface MESFET. The role of the contacts must clearly be taken into account.

(iv) For the hydrogen saturated surface we determined for the first time a value for the negative electron affinity of $\chi = -1.27$ eV. We have demonstrated a relationship between hydrogen coverage and electron affinity that is based on a simple surface dipole layer model. The electron affinity remains negative down to about 10% of a monolayer and saturates at +0.38 eV for the hydrogen-free 2×1 surface. Using χ as a measure for hydrogen coverage the kinetic parameters for the desorption process with an activation energy of 1.25 eV and a prefactor of 1×10^3 s^{-1} have been determined. As an additional piece of information to be elaborated upon elsewhere we find that the reconstruction sets in only after at least 70% of the hydrogen have been desorbed. This implies that an intermediate phase with a high coverage of dangling bonds is expected to exist at the (111) surface for a temperature of about 800°C which happens to be close to the optimum growth temperature of CVD diamond.

Acknowledgements

This work was supported by the Deutsche Forschungsgemeinschaft in the framework of the trinational 'D-A-CH' cooperation.

References

[1] Matsumoto S, Sato Y, Kamo M, Setaka S. Jpn J Appl Phys 1982;21:L183.

[2] Stöckel R, Stammler M, Janischowsky K, Ley L, Albrecht M, Strunk HP. J Appl Phys 1998;83:531.

[3] Ebert W, Vescan A, Gluche P, Borst T, Kohn E. Diam Rel Mat 1997;6:329.

[4] Kawarada H. Surf Sci Rep 1996;26:205.

[5] Gluche P, Aleskov A, Vescan A, Ebert W, Kohn E. IEEE Electr Dev Lett 1997;18:547.

[6] Himpsel FJ, Knapp JA, Van Vechten JA, Eastman DE. Phys Rev 1979;B20:624.

[7] Cui JB, Ristein J, Ley L. Phys Rev Lett 1998;81:429.

[8] Graupner R, Hollering M, Ziegler A, Ristein J, Ley L. Phys Rev 1997;B55:1084.

[9] Diederich L, Aebi P, Küttel O, Maillard-Schaller E, Fasel R, Schlapbach L. Surf Sci 1997;393:L77.

[10] Cui JB, Amtmann K, Ristein J, Ley L. J Appl Phys 1998;83:7929.

[11] Graupner R, Maier F, Ristein J, Ley L, Jung C. Phys Rev B 1998;B57:12397.

[12] Pandy KC. Phys Rev 1982;B25:4338.

[13] R. Graupner, Thesis, University Erlangen, 1997.

[14] Kern G, Hafner J, Kresse G. Surf Sci 1996;366:445.

[15] Ristein J, Stein W, Ley L. Phys Rev Lett 1997;78:1803.

[16] Rutter MJ, Robertson J. Phys Rev 1998;B57:9241.

[17] Bandis C, Pate BB. Phys Rev 1995;B52:12056.

[18] Bandis C, Pate BB. Surf Sci 1986;165:83.

[19] Cui, JB, Ristein, J, Ley, L, Phys Rev B, in press.

[20] Cui, JB, Graupner, R, Ristein, J, Ley, L, Diamond 98, in press.

[21] Topping J. Proc Roy Soc London 1927;A114:67.

[22] Values of α for CH_2 (1.64×10^{-40} Asm^2/V) and CH_3 (1.88×10^{-40} Asm^2/V) have been calculated by Gough, KM, J Chem Phys 1989;91:2424.

[23] Mönch, W, Semiconductor Surfaces and Interfaces, 2nd ed., Springer, 1995.

[24] Schulberg MT, Kubiak GD, Stulen RH. Mater Res Symp Proc 1992;270:401.

[25] Nishimori T, Sakamoto H, Takahuwa Y, Kono S. J Vac Sci Technol 1995;A13:2781.

Pergamon

Carbon 37 (1999) 801–805

CARBON

Mechanisms of surface conductivity in thin film diamond: Application to high performance devices

Hui Jin Looi[a], Lisa Y.S. Pang[a], Andrew B. Molloy[b], Frances Jones[b],
Michael D. Whitfield[a], John S. Foord[b], Richard B. Jackman[a],*

[a]*Electronic and Electrical Engineering, University College London, Torrington Place, London, WC1E 7JE, UK*
[b]*Physical and Theoretical Chemistry Laboratory, University of Oxford, South Parks Road, Oxford, OX1 3QZ, UK*

Received 17 June 1998; accepted 3 October 1998

Abstract

It has been known for some time that hydrogen within the bulk of diamond increases the conductivity of the material. However, only recently did it become apparent that the surface of thin film diamond can display p-type conductivity and that this too related to the presence of hydrogen. The origin of this effect has been controversial. We have used a wide range of techniques to study hydrogenated polycrystalline CVD diamond films to solve this problem. The generation of near surface carriers by hydrogen, which resides within the top 20 nm of 'as-grown' CVD films, is the origin of the conductivity rather than surface band bending which had also been proposed. Up to 10^{19} holes cm^{-3} can be measured and mobilities as high as 70 cm^2/Vs recorded. H-termination of the surface is important for the formation of high quality metal–diamond interfaces. © 1999 Elsevier Science Ltd. All rights reserved.

Keywords: A. Diamond; D. Electronic properties

1. Introduction

Thin film diamond, grown by chemical vapour deposition (CVD) techniques, typically displays p-type surface conductivity when in an 'as-grown' condition. Single crystal and polycrystalline films display this effect which appears to be lost following oxidation of the material leaving the surface highly resistive [1–9]. However, the origin of this effect remains controversial. A number of reports [1–3] propose that the carriers arise though the formation of shallow acceptor states beneath the surface when hydrogen is present. Other studies [4,5] suggest that it is surface bound hydrogen which causes band bending leading to an accumulation of holes in this region. In the case of polycrystalline diamond films, grown by chemical vapour deposition (CVD) methods, several groups [6–8] have also claimed that changes to the sp^2 (non-diamond) carbon may be responsible for variations in surface conductivity, rather than the presence of hydrogen itself. To complicate the picture further, in the absence of surface

hydrogen metal–diamond interfaces suffer from a 'pinned' Fermi level leading to Schottky barrier heights which are insensitive to the metal used; when hydrogenated this pinning effect is removed and near ideal interfacial characteristics have been measured [10–12].

Despite this confusion Hokazono and co-workers [13] have used hydrogenated single crystal diamond displaying p-type conductivity to fabricate very effective transistor structures. We have recently demonstrated that polycrystalline CVD diamond can be used in a similar manner [14]. It is therefore essential that a clearer understanding of the origin of the carriers and the electrical characteristics of hydrogenated diamond surfaces is achieved. To this end we have studied polycrystalline CVD diamond surfaces using Hall effect, secondary ion mass spectrometry (SIMS), X-ray and ultra-violet photoelectron spectroscopies (XPS, UPS) as well as current–voltage (I–V) measurements.

2. Experimental

Polycrystalline diamond, free standing grown by the

---

*Corresponding author.
E-mail address: r.jackman@eleceng.ucl.ac.uk (R.B. Jackman)

technique of microwave plasma enhanced CVD, was used throughout. Following growth, samples were cooled whilst being exposed to a pure hydrogen plasma. We have previously shown that the 'as-grown' surface of this material displays p-type conductivity whilst surfaces that are subsequently oxidised are highly resistive [9]. Films, which displayed an irregular grain structure with top side grain sizes in the range 20–40 μm, revealed an intense 1332 cm^{-1} when inspected with Raman spectroscopy, indicative of good quality CVD diamond [15]. Sheet resistance, carrier concentrations and carrier mobilities were assessed using Hall effect measurements made at room temperature (BioRad HL5200 with a magnetic field strength of 0.32 T). HP semiconductor parameter analysers 4145B and 4061A were used to probe the I–V characteristics of samples with evaporated metal contacts. Photoelectron spectra were recorded using a VG ESCALAB fitted with Al Ka (1486.6 eV) and Helium (I) (21.2 eV) X-ray and UV photon sources; a thin gold layer evaporated near the edge of each sample was used for calibration purposes. SIMS measurements were made by bombarding the surface with O_2^+ primary ions with an energy of 15 keV.

3. Results

Gold contacts on the 'as-grown' films gave near linear I–V plots, indicative of 'ohmic' behaviour. The room temperature I–V characteristics of adjacent contacts were investigated following heating. Experiments were performed in air and in-vacuo (10^{-7} mbar) and, interestingly, gave identical results. Fig. 1(a) shows the resistance measured for the film following heating to 60, 150 and 200°C for increasing time periods. Little change is evident at 60°C, but higher temperatures led to an increase in film resistance which saturates after a short period at a given temperature, increasing again at a higher temperature. Films were immersed both in water and acidic solutions, dried and then re-tested; no changes to the measured I–V characteristics could be provoked in this manner.

Figure 1(b) shows the change in sheet carrier concentration measured by Hall effect under similar conditions; annealing reduces the carrier concentration measured, again with stable values being achieved for each annealing temperature after a period of time. Fig. 2(a) reveals the Hall mobility value recorded, at room temperature, following annealing at 200°C for increasing periods; the value reaches 70 cm^2/Vs for samples heated for periods longer than 160 min.

As-grown samples were analysed using SIMS and compared to heat treated films (200°C in air for a sufficient period to stabilise the I–V characteristics of the film). The SIMS signals recorded at 1amu (H) are plotted in Fig. 2(b) as a function of sputter depth. The as-grown film (trace (i)) produces a strong hydrogen signal over the first 20 nm of diamond with a significant tail persisting to depths greater

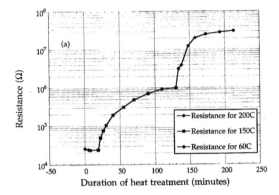

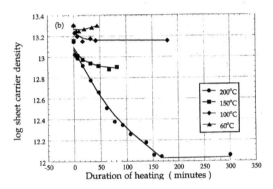

Fig. 1. (a) Room temperature resistance measured between adjacent Au contacts plotted as a function of time that the film had been previously heated to various temperatures; (b) Sheet carrier concentration measured at room temperature using the Hall effect method as a function of the time that the film had been previously heated at various temperatures.

than 60 nm; this tail may be due to re-adsorption of sputtered hydrogen. In contrast the heated film reveals little hydrogen presence at depths greater than around 10 nm.

UPS data recorded with He(I) radiation is shown in Fig. 3(a). The as-grown sample shows a broad band centred at 15 eV with a slight shoulder at ~17eV. A more pronounced shoulder on the broad band is evident at 9.5 eV. Upon heating the band narrows by ~0.2 eV and shifts to lower kinetic energies, becoming centred at ~14 eV. The 9.5 eV shoulder does not shift in energy but does reduce in intensity. X-ray photoelectron spectra were recorded in the C 1s and O 1s binding energy regions again for the as-grown material and samples that had been heated to 60 and 200°C for differing periods. Fig. 3(b) shows the C 1s peak to be at an energy of 285.5 eV on the as-grown sample; the peak broadens upon heating and a second peak, located at ~286.2 eV can be identified. A weak O 1s peak is also visible on the as-grown sample, located at 534

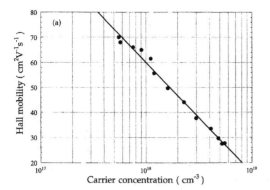

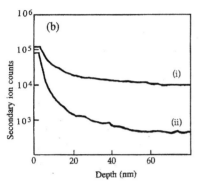

Fig. 2. (a) Hall mobility values recorded for the films shown in 1(b); (b) SIMS 1amu (H) signal recorded for (i) an 'as-grown' diamond film and (ii) a film that has been heated to 200°C for 120 min.

eV; heating in air has little effect on the O/C peak ratio which remains at ~0.1.

4. Discussion

The Au–diamond Schottky barrier height would be close to zero in an ideal case, which explains the near ohmic characteristics measured here. These values indicate that the surface is relatively free from interface states which would pin the Fermi level, as has been observed on hydrogenated single crystal diamond [10–14]. The dependence of room temperature conductivity on heat treatments given to hydrogenated films (Fig. 1(a)) has been observed previously. For example, Muto and co-workers [7] felt that the resistivity increase they measured when thin film (1 μm) hydrogenated polycrystalline CVD diamond was heated was due to the removal of disorded graphite. The increase in resistance measured as films are heated (Fig. 1(a)) can be caused by a loss of carriers within the film and/or an increase in the specific contact resistance.

However, the measurements using Hall effect apparatus, where the effect of specific contact resistance changes will be ignored (Fig. 1(b)), show that annealing in air at temperatures up to 150°C increases the sheet resistivity and decreases the sheet carrier concentration. After some period of time all measured values remain constant at a given temperature. The increase in resistivity after annealing can either be caused by the loss of carriers from the surface or the generation of more scattering centres. Since the mobility is not significantly altered the former case appears more likely. Annealing at 200°C gives a slightly different result; sheet resistivity and carrier concentrations increase and decrease as before, but Hall mobility values rise after a (small) initial fall.

Hall effect measurements on the p-type conductive layer that results from hydrogenation of diamond have previously been confined to homoepitaxial films. Hayashi and co-workers [4] found that as-deposited films supported carrier densities some 4–5 orders of magnitude higher than oxidised films. SIMS analysis revealed a hydrogenated layer around 20 nm thick which was removed upon oxidation; p-type conduction (4×10^{18} cm^{-3}, 30 cm^2/Vs) was attributed to shallow acceptor levels generated by the hydrogen. In the current study SIMS revealed a near surface hydrogen rich layer within the diamond over a similar depth. Heating reduced both the depth and magnitude of this layer. Although a quantitative assessment can not be made, this trend correlates with the reduction in carrier concentration that we have observed. It is apparent from the present study that carrier concentration can be controlled by annealing alone; surface oxidation is not necessary.

Adsorbed hydrogen is not expected to be thermally desorbed at the temperatures investigated here [16,17]; this assertion is supported by the absence of significant new structure in the UPS spectrum as the sample is heated (Fig. 3(a)). The UPS peak at 9.5 eV is due to secondary electrons; the samples were not electrically biased during UPS measurements leading to the absence of any NEA peak in the spectra presented [18]; a detailed analysis on the NEA properties of these samples will be published elsewhere. The primary C 1s XPS peak does not shift upon heating indicating that significant changes to any band bending cannot be taking place. The appearance of a high binding energy component to the C 1s peak can be attributed to reacted C forms, such as C–H [19]. C 1s peak shifts have been observed by others [5,20,21] when hydrogenated surfaces are oxidised; changes to the extent of band bending were thought responsible.

Shirafuji and Sugino [5] have presented a detailed study on the electrical properties of diamond surfaces which concluded that surface band bending was the origin of p-type conduction in hydrogenated diamond films. Hydrogenated surfaces showed upward band bending which would enable hole accumulation at the surface to occur. Downward band bending was promoted by oxidation but

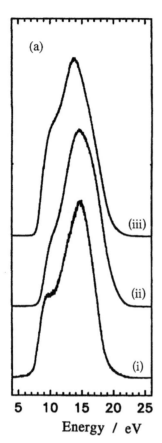
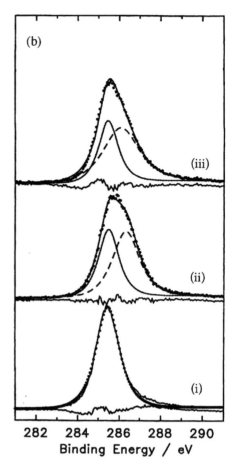

Fig. 3. (a) UPS data recorded for (i) an 'as-grown' diamond film and (ii) a film that has been heated to 200°C for 120 min; (b) XPS data recorded for (i) an 'as-grown' diamond film and (ii) a film that has been heated to 200°C for 120 min.

by an insufficient amount to cause electron accumulation. However their experiments did not deconvolute the possible effects of heat treatments and oxidation. The measurements presented above clearly show significant changes in carrier concentration and film resistivity occur under conditions where oxidation and band bending is not apparent. Thus, whilst band bending and hole accumulation may be occurring on hydrogenated surfaces this is not the origin of the high carrier concentrations and resistivity changes measured here. Film characteristics are not effected by immersion in water or acid in the absence of heating suggesting that the model involving acidic surface charge transfer reactions proposed by Gi et al. [22] can not explain the conductivity we have observed.

Our results are consistent with the model proposed by Hayashi and co-workers [12]. They observed a near

surface hydrogen layer in SIMS and were able to associate this phase with a peak at 540 nm in cathodoluminscence studies; this implies that the hydrogen was promoting the formation of gap states, some of which act as shallow acceptors leading to p-type diamond. The correlation found here in film resistivity and carrier concentration with a reduction in near surface hydrogen in the absence of band bending changes would appear to confirm that this is the primary origin of the hole concentrations observed.

5. Concluding remarks

A wide range of scientific techniques have been used simultaneously to study the origin of p-type conduction in diamond films. As-grown polycrystalline CVD material has a relatively high near surface concentration of hydrogen, presumably due to the way that CVD growth is typically terminated. Heating the film to 200°C causes some of this hydrogen to be lost; this process provokes a reduction in the hole conductivity within the film. The *surface* remains hydrogenated and no change to any surface band bending present is detected. In previous studies the modification of near surface hydrogen profiles and surface oxidation were often interlinked processes, making it difficult to establish the origin of p-type conductivity. The current study shows that it is hydrogen within diamond that gives rise to carriers; shallow acceptor states, and possibly other deeper gap states, must form when hydrogen is within the diamond lattice in these concentrations. However, H-terminated surfaces are critically important in forming good quality contacts and for achieving properties such as NEA. The exact nature of the hydrogen promoted shallow acceptor states remains unknown and is worthy of further investigation.

References

[1] Shiomi H, Nishibayashi Y, Fujimori N. Jpn J Appl Phys 1991;30:1363.

[2] Maki T, Shikama S, Komori M, Sakaguchi Y, Sakuta K, Kobayashi T. Jpn J Appl Phys 1992;31:1363.

[3] Hayashi K, Yamanaka S, Okushi H, Kajimura K. Appl Phys Lett 1996;68:376.

[4] Shirafuji J, Sugino T. Diam & Relat Mater 1996;5:706.

[5] Kawarada H, Sasaki H, Sato A. Phys Rev 1995;B 52:11351.

[6] Mori Y, Show Y, Deguchi M, et al. Jpn J Appl Phys 1993;32:L987.

[7] Muto Y, Sugino T, Shirafuji J, Kobashi K. Appl Phys Lett 1991;59:843.

[8] Kulkarni AK, Shrotriya A, Cheng P, Rodrigo H, Bashyam R, Keeble DJ. Thin Solid Films 1994;253:141.

[9] Looi HJ, Jackman RB, Foord JS. Appl Phys Lett 1998;72:353.

[10] Kawarada H, Aoki M, Sasaki H, Tsugawa T. Diam. & Relat Mater 1994 961.

[11] Kiyota H, Okushi H, Ando T, Kamo M, Sato Y. Diam & Relat Mater 1996;5:718.

[12] Hayashi K, Watanabe H, Yamanaka S, Sekiguchi T, Okushi H, Kajimura K. Diam & Relat Mater 1997;6:303.

[13] Hokazono A, Ishikura T, Nakamura K, Yamashita S, Kawarada H. Diam & Relat Mater 1997;6:339.

[14] Looi HJ, Pang LYS, Wang Y, Whitfield MD, Jackman RB. IEEE Electron Device Lett 1998;19:112.

[15] Grot SA, Gildenblat GSh, Hatfield CW, et al. IEEE Electron Device Lett 1990;11:100.

[16] Jackman RB, Chua LH, Foord JS. Surface Sci 1993;292:47.

[17] Chua LH, Jackman RB, Foord JS, Chalker PR, Johnston C, Romani S. J Vac Sci & Tech 1994;12:3033.

[18] McKeag RD, Jackman RB. Diamond & Relat Mater 1998;7:513.

[19] Diederich L, Kuttel OM, Schaller E, Schlapbach L. Surface Sci 1996;349:176.

[20] Haasz K. J Vac Sci Tech A 1996;14:189.

[21] Bachmann PK, Eberhardt W, Kessler B, et al. Diam & Relat Mater 1996;5:1378.

[22] Gi RS, Mizumasa T, Akiba Y, Hirose Y, Kurosu T, Iida M. Jpn J Appl Phys 1995;34:5550.

Pergamon

Carbon 37 (1999) 807–810

CARBON

A large range of boron doping with low compensation ratio for homoepitaxial diamond films

J.-P. Lagrange, A. Deneuville*, E. Gheeraert

Laboratoire d'Etudes des Propriétés Electroniques des Solides, Centre National de la Recherche Scientifique BP 166, 38042 Grenoble Cedex 9, France

Received 16 June 1998; accepted 3 October 1998

Abstract

A large range ([B] from 5×10^{16} to 8×10^{20} cm^{-3}) of boron concentration has been obtained for homoepitaxial diamond films. From the variation of the conductivity versus temperature between 300 and 1000 K, a simpler method is used to derive the activation energy(ies) E_a and the compensation ratio than from the variation of the hole concentration and mobility with temperature. For [B]$<2 \times 10^{17}$ cm^{-3}, a saturation of the conductivity is observed between 680 and 1000 K, $E_a = E_i$ (ionisation energy of the boron), and compensation ratio around 10% is deduced. Then, up to 10^{19} cm^{-3}, an additional $E_a = E_i/2 = 185$ meV is observed between 500 and 1000 K, with compensation ratio <10%. At higher [B], an additional nearest neighbour hopping contribution to the conduction is obtained from the formation of the boron impurity band. It dominates and gives metallic conductivity when [B]$=3 \times 10^{20}$ cm^{-3}. © 1999 Elsevier Science Ltd. All rights reserved.

Keywords: A. Diamond; D. Electronic properties

1. Introduction

From their exceptional electronic and thermal properties [1], boron-doped diamond thin films appear well-suited for power and high-temperature operating devices. However, this requires the control of both high (ohmic contacts) and low doping levels with the achievement of a low concentration of residual defects (efficient space charge zone) as some of them act as compensating donor levels. Previous work [2–14] measured the temperature variation of the conductivity, and sometimes that of the hole concentration and of the Hall mobility, but only some works [2,8–12] studied them from the compensation point of view. On the other hand, the lowest boron concentration exhibiting the usual doping effect is an upper limit to the residual donors.

From the early works [3–6], the lowest efficient boron concentration level in films has improved from 4×10^{17} [4] to about 1.8×10^{16} cm^{-3} [2] (the lowest boron concentration in the best natural IIb crystal [15,16]). From the similarity with the results obtained for the best natural

crystals, the boron ionisation energy (E_i around 0.37 eV) as the activation energy E_a for the conductivity and the carrier concentration was considered as a criterion for perfect samples in the early works, while this corresponds to a significant compensation ratio ($K>10$%), as we will see in Section 3.2. Some recent works [2,8–12] now look at K, with usually $15<K<25$% [9–12], but their temperature range is sometimes too restricted to extract the full information from this type of measurement. They report K around 10% for [B] in the range 10^{16} cm^{-3} [2,11] and in the range 2 to 10% for [B] in the range 10^{18} cm^{-3} [8–10] in some exceptional ('selected') samples.

We show here, that a simple method can be used which extracts the B ionisation energy and the compensation ratio K of these films from the variation of their conductivity versus temperature up to 1000 K.

2. Experimental details

Homoepitaxial boron-doped diamond films were grown by Microwave Plasma-assisted Chemical Vapor Deposition, on (100) type Ib synthetic diamond at 820°C, 30 Torr, 4% of methane with some diborane (from [B]/[C]=0.25 to [B]/[C]=4000 ppm) and 96% of hydrogen. A buffer layer

---

*Corresponding author. Fax: +33-04-7688-7988.
E-mail address: deneuvil@lepes.polycnrs-gre.fr (A. Deneuville)

of 0.5 μm of undoped diamond was grown before the 3-μm thick doped film.

The room temperature resistivities were measured in a four-points probe configuration with 1-mm spacing.

Measurements were done from 300 to 1000 K, of the conductivity (in coplanar geometry with two tungsten tips, around 1 mm spaced) for all samples, and from 300 to 850°K of the hole concentration and mobility on one sample to verify the validity of our interpretation of $\sigma(T)$.

As the ionisation ratio of the boron atoms was very low at room temperature, their concentrations in the films were measured by infrared absorption spectroscopy between the fundamental and the excited levels of the bound hole [17] after calibration by SIMS. The linear relationship between the boron concentration in the solid phase and in the gas phase found between $[B]=6\times10^{17}$ and 3×10^{20} cm^{-3} was extrapolated from 5×10^{16} to 8×10^{20} cm^{-3}.

3. Results and discussion

3.1. Resistivity vs. [B]

Figure 1 plots the room temperature resistivities of the diamond films vs. the boron concentration. It decreases by eight orders of magnitude (10^5 to 10^{-3} Ω cm) for boron concentration of 2×10^{16} to 8×10^{20} cm^{-3}. The two lower values correspond to a metallic conductivity as will be shown later, and the variation is slow with the boron content. On the contrary, we do not see saturation towards the lower boron content. The resistivity of our undoped films was around 10^{13} Ω cm, therefore we expect a larger range of resistivity controlled by the boron content (see

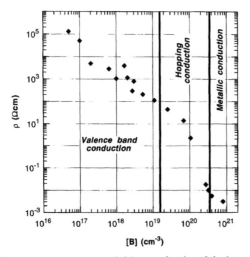

Fig. 1. Room temperature resistivity as a function of the boron concentration.

Section 3.2) before pinning of the Fermi level by the residual defects if boron incorporation lower than 2×10^{16} B cm^{-3} can be controlled. Finally there is fast variation of the resistivity from 10^2 to 10^{-2} Ω cm for boron concentration of 2×10^{19} to 3×10^{20} cm^{-3}, which is ascribed to an additional hopping component between unionized and ionized boron levels (see Section 3.2).

3.2. Electrical measurements vs. temperature

The theoretical expression of the conductivity in a compensated semiconductor [18] in the ionisation regime is:

$$\frac{1}{\sigma}=\frac{1}{\sigma_1}+\frac{1}{\sigma_2}$$

$$\sigma_1=q\mu\frac{N_V}{g_a}\left(\frac{N_A-N_D}{N_D}\right)\exp\left(-\frac{E_i}{kT}\right)$$

$$\text{with}\begin{cases}\mu=\mu_{300}\left(\dfrac{T}{300}\right)^{-s}\\ N_V=N_0T^{3/2}\end{cases}$$

$$\sigma_2=q\mu\left(\frac{N_V}{g_a}(N_A-N_D)\right)^{1/2}\exp\left(-\frac{E_i}{2kT}\right) \quad(1)$$

where N_A and N_D are respectively the acceptor and the compensating donor concentration, N_V the density of state of the valence band, E_i the ionisation energy of the acceptor, g_a the degenerescence level of the acceptor state ($g_a=4$), and s the exponent of the usual power law variation of the mobility. σ_1 is the conductivity for low temperature or high degrees of compensation (corresponding to $p_1\ll N_D$). σ_2 is the conductivity in an uncompensated semiconductor (corresponding to $p_2\gg N_D$), the corresponding hole concentrations in the valence band p (p_1 and p_2) can be deduced using $\sigma=q\mu p$. A compensation ratio lower than 10% can be deduced unambiguously only when p and $\sigma(T)$ curves exhibit above room temperature two activation energies (E_i at 'low' temperature and $E_i/2$ at 'high' temperature) instead of one (E_i). On the other hand, saturation of p or σ below 1000 K can be obtained only when $[B]<10^{17}$ cm^{-3} with compensation ratio $\leq10\%$.

Figures 2 and 3 show respectively the hole concentration and the mobility versus the reciprocal temperature for a particular sample. The activation energy varies between E_i and $E_i/2$, which indicates a low compensation ratio (<10%). We have used $N_0=3.4\times10^{15}$ cm^{-3} K$^{-3/2}$. The fit of the curve gives respectively 7×10^{17} and 4.9×10^{16} cm^{-3} for N_A and N_D and therefore a compensation ratio of 7%. Consistently, a hole concentration near the N_A-N_D is obtained at 850°K. On the same sample (Fig. 3), $\mu=\mu_{300}$ $(T/300)^{-2.6}$ with $\mu_{300}=553$ cm^2 V^{-1} s^{-1} on a large range of temperature. Quite similar values of the activation energy and of the compensation are obtained from the

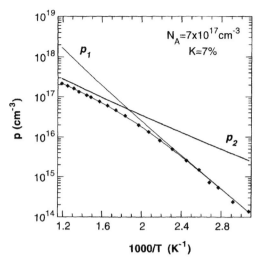

Fig. 2. Hole concentration versus reciprocal temperature for a particular sample. Fit in a compensated semiconductor model.

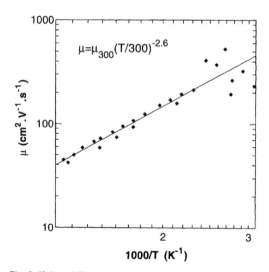

Fig. 3. Hole mobility versus reciprocal temperature for the same particular sample.

$\sigma(T)$ curve of the same sample using $\mu = \mu_{300}$ $(T/300)^{-2.6}$, therefore similar $\mu(T)$ will be used for the other samples (Fig. 4).

For the lower $[B] < 6 \times 10^{17}$ cm^{-3}, $E_a = E_i$ around 0.37 eV, the compensation ratio is over 10%. On the other hand the saturation of the conductivity around 680 K obtained here when $[B] \leq 1 \times 10^{17}$ cm^{-3} confirms these low boron incorporation levels and indicates a compensation ratio around 10%. In the medium range, up to about $[B] = 10^{19}$

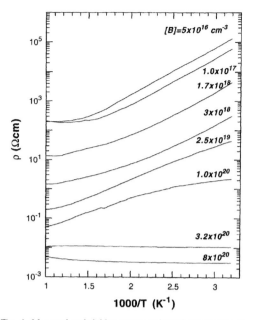

Fig. 4. Measured resistivities versus reciprocal temperature for some samples.

cm^{-3}, two conductivity components with activation energies E_i and $E_i/2$ with E_i around 0.37 eV are required to fit the experimental curves. Compensation ratios are determined between 2 and 7%. In both ranges, we obtain for all samples the best K values obtained only previously for some selected samples [2,8–11].

Above about 10^{19} B cm^{-3} an additional conduction process appears in the lower temperature range, and increases as the boron content increases. It is due to nearest neighbour hopping (NNH) between ionised and neutral acceptors within the boron impurity band as shown by the analysis of $\sigma(T)$ between 100–500 K. Its [B] threshold concentration agrees with the 10^{19} cm^{-3} value predicted by Werner et al. [14].

For $[B] = 2.7 \times 10^{20}$ or higher, the conductivity decreases with the temperature, (metallic behaviour), in accordance with the Mott transition expected around 2×10^{20} cm^{-3} [19].

3.3. Activation energy vs. [B]

From the previous analysis, the boron ionisation energy E_i remains around 0.37 eV at low [B], decreases for $[B] > 3 \times 10^{18}$ cm^{-3}, and falls to zero for $[B] = 10^{20}$ cm^{-3}. This evolution is in qualitative agreement with the previous measurements [8,12,20,21] as shown on the Fig. 5. Lee and McGill [22] have shown for silicon that the decrease in the ionisation energy in not directly related to the

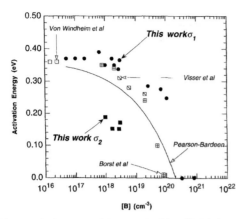

Fig. 5. Activation energy of the σ_1 conductivity (filled circles) and of the σ_2 conductivity (filled squares) as a function of boron incorporation. The evolution predicted by the model of Pearson and Bardeen [23] is plotted only for orientation. Some values reported in the literature are plotted for comparison [8,12,21] assuming $[B] \approx N_A - N_D$.

majority impurity concentration (Pearson–Bardeen model [23], full line on Fig. 5), but depends also on the compensation ratio (and will be constant with a zero compensation ratio). In boron-doped synthetic diamond, Bourgoin et al. [20] found a good correlation between the ionisation energy and the concentration of the compensating donors. A similar conclusion has been obtained for GaP and ZnSiP$_2$ by Monecke et al. [24]. Our nearly constant values of E_i suggest here that the compensation ratio remains weak (lower than or around 10%) up to $[B] = 3 \times 10^{18}$ cm^{-3}, then decreases near the formation of the boron impurity band above about 10^{19} B cm^{-3}.

4. Conclusion

We have grown homoepitaxial diamond thin films boron-doped from 5×10^{16} to 8×10^{20} cm^{-3}. The $I(T)$ curves have been fitted by the general expression of the conductivity in a compensated semiconductor, with low compensation ratio. For $[B] < 2 \times 10^{17}$ cm^{-3}, the saturation of the conductivity is observed between 680 and 1000 K and the compensation ratio is around 10%. Then up to 10^{19} cm^{-3}, besides the high activation energy around E_i at 'low' temperature, an activation energy of half the boron ionisation energy E_i is observed at high temperature and compensation ratio < 10% are obtained. We obtain for all

samples in these two [B] ranges the best K of some selected samples of previous works. For higher B incorporation, there is a decrease of the activation energy near the formation of the boron impurity band which supplies an additional near neighbour hopping conduction at low temperature. For $[B] = 3 \times 10^{20}$ cm^{-3}, there is metallic conduction on the boron impurity band.

References

[1] Gildenblat GS, Grot SA, Badzian A. Proc IEE 1991;79:647.
[2] Fox BA, Hartsell ML, Malta DM, Wynands HA, Tessmer GJ, Dreyfus DL. Proc Mat Res Soc Symp 1996;146:319.
[3] Fujimori N, Imai T, Doi A. Vaccum 1986;6:99.
[4] Okano K, Naruki H, Akiba Y, Kurosu T, Iida M, Hirose Y, et al. Jap J Appl Phys 1989;28:1066.
[5] Fujimori N, Nakahata H, Imai T. Jap J Appl Phys 1990;29:824.
[6] Shiomi H, Nishibayashi Y, Fujimori N. Jap J Appl Phys 1991;30:1363.
[7] Grot SA, Hatfield CW, Gildenblat GSh, Badzian AR, Badzian T. Appl Phys Lett 1991;58:1542.
[8] Visser EP, Bauhis GJ, Janssen G, Vollenberg W, van Enckevort WJP, Gilling LJ. J Phys: Condens Mater 1992;4:7365.
[9] Malta DM, von Windheim JA, Fox BA. Appl Phys Lett 1993;62:2696.
[10] Malta DM, von Windheim JA, Wynands HA, Fox BA. J Appl Phys 1995;77:1536.
[11] Fox BA, Hartsell ML, Malta DM, Weynands HA, Kao CT, Plano LS, et al. Diamond Relat Mater 1995;4:622.
[12] Borst TH, Weis O. Diamond and Relat Mater 1995;4:948.
[13] Collins AT, Williams AWS. J Phys C: Solid State Phys 1971;4:1789.
[14] Werner M, Locher R, Kolhy W, Holmes DS, Klose S, Fecht HJ. Diamond and Relat Mater 1997;6:308.
[15] Williams AWS. Ph.D. thesis. University of London 1971.
[16] Collins AT. In: Field JE, editor. The properties of diamond. Academic Press, 1979.
[17] Gheeraert E, Deneuville A, Mambou J. Diamond and Relat Mater 1998;7:1509.
[18] Blakemore. Semiconductor statistics. New York: Dover Publications, Inc., 1987.
[19] Williams AS, Lightowlers EC, Collins AT. J Phys C: Solid State Phys 1970;3:1727.
[20] Bourgoin JC, Krynicki J, Blanchard B. Phys Stat Sol (a) 1979;52:293.
[21] von Windheim JA, Venkatesan V, Malta DM, Das K. J Electron Mater 1993;22:391.
[22] Lee TF, McGill TC. J Appl Phys 1975;46:373.
[23] Pearson GL, Bardeen J. Phys Rev 1949;75:865.
[24] Monecke J, Diegel W, Ziegler E, Kuhnel G. Phys Stat Sol (b) 1981;103:269.

Pergamon

Carbon 37 (1999) 811–816

CARBON

Minority-carrier transport parameters in CVD diamond

S. Salvatori*, M. C. Rossi, F. Galluzzi

Dipartimento di Ingegneria Elettronica and INFM, Università degli Studi Roma Tre, V. Vasca Navale 84, I-00146 Roma, Italy

Received 16 June 1998; accepted 3 October 1998

Abstract

A study is presented on minority carrier transport parameters, in particular mobility–lifetime products, in natural and CVD diamond materials with different morphology. $\mu\tau$ products are obtained by photocurrent versus field measurements and their values range between 10^{-8}–10^{-5} cm^2 V^{-1} depending on grain size and orientation. Observed variations are mainly related to mobility changes, while minority carrier lifetimes are almost constant around 10^{-10}–10^{-9} s. The origin of such a behaviour is analysed in terms of gap state distribution arising from structural defects, impurities and non-diamond microphases. © 1999 Elsevier Science Ltd. All rights reserved.

Keywords: A. Diamond; B. Chemical vapor deposition; D. Photoconductivity

1. Introduction

The availability of Chemical Vapor Deposition (CVD) techniques for diamond film growth on silicon substrates recently induced a strong interest in electronic applications of diamond and provoked much work on transport mechanisms in such wide band-gap semiconductors. However minority carrier transport properties have been investigated to a lesser extent, in spite of their relevance for optoelectronic applications. Here we present an experimental study on this latter aspect, analysing in particular the values of minority carrier mobility–lifetime products ($\mu\tau$) in diamond materials with different morphologies.

Indeed transport properties of minority charge carriers can be adequately summarised by the $\mu\tau$ product, which defines both the diffusion length $L_{df} = (\mu\tau kT/q)^{1/2}$ and the drift length $L_{dr} = \mu\tau E$ (here k is the Boltzmann constant, q the electron charge, T the absolute temperature and E the applied electric field). $\mu\tau$ values, which can be evaluated either by separate measurements of mobility and lifetime or -as in the present case-by photocurrent versus field measurements, are strongly dependent on semiconductor structure and purity. For example, in single crystal silicon, electrons can exhibit $\mu\tau$ values as high as 0.1–1.0 cm^2 V^{-1}, which drop to about 10^{-4} cm^2 V^{-1} in "solar

grade" polycrystalline silicon and to about 10^{-8} cm^2 V^{-1} in hydrogenated amorphous silicon.

2. Experimental

Polycrystalline diamond films have been grown by using microwave plasma-assisted and hot filament chemical vapor deposition techniques (MW-PACVD and HFCVD, respectively). Details on the deposition systems are described elsewhere [1,2]. The morphology of the films, observed by means of scanning electron microscopy (SEM), can be divided in small-grained (1–2 μm) randomly oriented films (HFCVD) and middle- or large-grained (5–20 μm) textured films (MW-PACVD).

Metal–semiconductor–metal (MSM) structures were prepared by evaporating 0.2 μm-thick silver or aluminium layer on diamond surface. Interdigitated contacts (with interelectrode distance of 10 μm) were obtained by using optical lithography and wet etching. An example of such a structure is reported in Fig. 1.

Steady-state photocurrent measurements were performed with a 300 W Xenon source lamp coupled to a Jobin Yvon H10 monochromator with a grating blazed at 250 nm and by using standard lock-in technique.

Transient photoresponse was measured by using VUV (125 nm) and UV (313 nm) pulsed lasers (FWHM≈7 ns) as light sources. VUV light was generated by four wave

---

*Corresponding author.

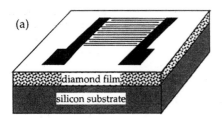

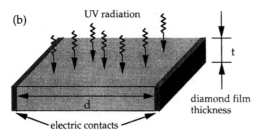

Fig. 1. (a) Experimental configuration of the diamond-based MSM structure; (b) simplified scheme of the MSM structure used for the theoretical models.

sum-frequency mixing in mercury vapor as described in details in ref. [3]. Photocurrent measurements were performed by using coaxial bias insertion tee (Picoseconds Pulse Labs, mod. 5531) and 1 GHz oscilloscope (Tektronix, mod. TDS540A).

3. Theoretical models

As mentioned above, minority carrier mobility–lifetime products in diamond can be obtained from steady-state measurements of photocurrent versus applied field strength. We consider in particular MSM structures (see Fig. 1) with either ohmic or rectifying behaviour.

In both cases the photocurrent can be expressed by the following relationship [4]:

$$I_{ph} = qF_0 f_{abs} \eta_{coll} \qquad (1)$$

where q is the electron charge, F_0 is the number of incident photons per unit time and η_{coll} is the charge-carrier collection efficiency, depending on the nature of the metal/diamond contact. The absorbed photon fraction can be written:

$$f_{abs} = (1 - R)(1 - e^{-\alpha t}) \qquad (2)$$

(here R and α are the reflectivity and the absorption coefficient of the diamond film and t is its thickness) which reduces to $f_{abs} \approx (1-R)\alpha t$ for near-gap UV illumination ($\lambda \approx 200$–230 nm).

In the following model we assume for simplicity an uniform electric field distribution. This assumption is expected to fail at low fields where non uniformities in both dimensions can actually appear. In this case field strength refers to a space-averaged value.

Assuming electrons as minority carriers (indeed non intentionally doped CVD diamond exhibits p-type behaviour), the collection efficiency in ohmic structures is [4]:

$$\eta_{coll}^{ohm} = (\nu_n + \nu_p)\tau_n/d = \nu_n \tau_n (1 + r)/d \qquad (3)$$

where ν_n, ν_p are electron and hole drift velocities, τ_n is electron lifetime, d the interelectrode distance and r is the ratio between hole and electron velocities.

In rectifying MSM structures, assuming a perfectly symmetric back-to-back diode configuration, the total photocurrent depends on the applied voltage V according to the equation:

$$I_{ph} = i_{ph} \, tgh\left(\frac{V}{2nV_T}\right) \qquad (4)$$

where n is the diode quality factor, V_T is the thermal voltage and i_{ph} is the photocurrent in a single Schottky diode. Neglecting for simplicity diffusion transport (indeed, as shown below, actual diffusion lengths in CVD diamond films are about 0.1–0.5 μm), i_{ph} can be expressed by the Hecht formula [5,6]:

$$i_{ph} = qF_0 f_{abs} \frac{\nu_n \tau_n}{d} (1 - e^{-d/\nu_n \tau_n}) \qquad (5)$$

so that the overall collection efficiency can be written in this case:

$$\eta_{coll}^{rect} = \frac{\nu_n \tau_n}{d} (1 - e^{-d/\nu_n \tau_n}) \, tgh\left(\frac{V}{2nV_T}\right) \qquad (6)$$

We have to remark that, in general, drift velocity is a non linear function of the electric field $\mathscr{E} \approx V/d$, according to the phenomenological relationship [7]:

$$\nu_{n(p)} = \frac{\mu_{n(p)} \mathscr{E}}{1 + \mu_{n(p)} \mathscr{E}/\nu_{n(p)}^{sat}} \qquad (7)$$

where $\mu_{n(p)}$ is the electron (hole) drift mobility and $\nu_{n(p)}^{sat}$ the electron (hole) saturation velocity. For single crystal type IIa natural diamond experimental mobility and saturation velocity values [8] are:

$$\mu_n \approx 2400 \text{ cm}^2\text{V}^{-1}\text{s}^{-1}; \ \mu_p \approx 2000 \text{ cm}^2\text{V}^{-1}\text{s}^{-1}$$

$$\nu_n^{sat} \approx 1.6 \cdot 10^7 \text{ cm s}^{-1}; \nu_p^{sat} \approx 1.1 \cdot 10^7 \text{cm s}^{-1} \qquad (8)$$

so that we have the velocity-field dependencies shown in Fig. 2 (curves a, b).

As a consequence of the slight difference between the electron and hole mobilities and saturation velocities, the

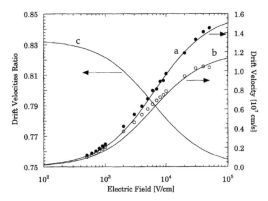

Fig. 2. Electron and hole drift velocities (a, b) and drift velocity ratio as a function of the applied field for single crystal natural diamond [4].

ratio r exhibits in natural diamond only a weak dependence on field strength, varying in the range 0.75–0.84, as shown in Fig. 2 (curve c).

From Eq. (3) and Eq. (6), at relatively low fields ($\mathscr{E} < \nu_n^{sat}/\mu_n$, $d/\mu_n\tau_n$ but $\mathscr{E} > 2nV_T/d$) both ohmic and rectifying MSM structures show a similar linear dependence between collection efficiency and field:

$$\eta_{coll} \approx K \frac{\mu_n\tau_n}{d} \mathscr{E} \qquad (9)$$

($K = 1$ for rectifying contacts and $K = 1 + r$ for ohmic contacts), whereas at very high fields ($\mathscr{E} >> \nu_n^{sat}/\mu_n$, $d/\mu_n\tau_n$) saturation occurs in both structures but at different values:

$$\eta_{sat}^{ohm} \approx \nu_n^{sat}\tau_n/d \qquad (10)$$

$$\eta_{sat}^{rect} \approx 1 \qquad (11)$$

As an example, calculated η_{coll} versus $\mathscr{E}$ plots for ohmic and rectifying MSM structures on single crystal natural diamond are shown in Fig. 3 assuming the data (8), $\tau_n = 1$ ns and $d = 10$ μm.

In conclusion photocurrent versus field measurements over a large field strength range allow to evaluate mobility–lifetime product (see Eq. (7) and –in the case of photoconductive devices– can also give a separate estimate of minority carrier lifetime (Eq. (9) and (Eq. (10))).

4. Experimental results and discussion

In order to check the validity of the method and to obtain a reference value for the mobility–lifetime product, first we analysed photocurrent-field measurements on a single crystal natural diamond with rectifying MSM structure, illuminated at 200 nm [9]. As shown in Fig. 4, best fit of experimental data with Eq. (6) gives a $\mu\tau$ product of the order of 10^{-5} cm^2 V^{-1}. Assuming a mobility value around 2400–2500 cm^2 V^{-1} s^{-1} we can deduce a minority carrier lifetime in the nanosecond range. This result compares fairly well with estimation of 740 ps reported by S. Han et al. [10] on the same data and with the value of several hundreds of ps obtained on similar materials by transient photoconductivity technique [10–12].

Photocurrent versus field measurements have been then performed on both rectifying and ohmic MSM devices based on thin diamond films grown by HFCVD and exhibiting a small-grain randomly oriented structure. Typical results for rectifying and ohmic contacts, under illumination in the 185–215 nm range, are reported in Figs. 5 and 6 respectively. In the former case the photocurrent tends to saturate at fields lower than 10^5 V cm^{-1}, while in

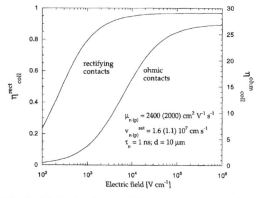

Fig. 3. Calculated collection efficiencies as a function of the applied field for rectifying and ohmic MSM structures based on single crystal natural diamond.

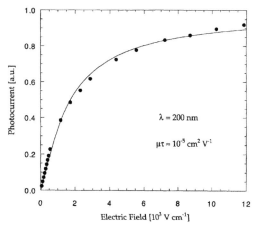

Fig. 4. Photocurrent versus applied field for rectifying MSM structure based on natural diamond IIa [5]. Best-fit curve according to Eq. (2), Eq. (6), Eq. (7).

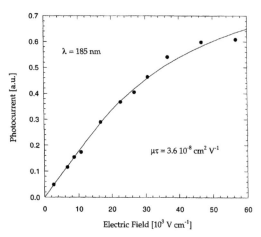

Fig. 5. Photocurrent versus applied field for rectifying MSM structure based on unoriented small-grain HFCVD diamond film. Best-fit curve as in Fig. 4.

the latter case a linear dependence on the applied field is observed up to $2 \cdot 10^5$ V cm^{-1}. This different behaviour reflects the different electrical structure, according to Eqs. (3) and (6), but in both cases the evaluated mobility–lifetime product is around 10^{-8} cm^2 V^{-1}. Mobility values typically reported for similar unoriented diamond films, although widely scattered, are around 50–100 cm^2 V^{-1} s^{-1}, much lower than single crystal mobilities due to the large scattering efficiency of grain boundaries and related microphases (in particular amorphous carbon). The measured $\mu\tau$ product should correspond therefore to a minority carrier lifetime of about 0.1 ns, i.e. only one order of magnitude lower than that estimated in single crystal diamond.

Finally I_{ph} versus $\mathscr{E}$ measurements under $\lambda = 215$ nm illumination, have been carried out on photoconductive MSM devices based on thick diamond films grown by MW-PACVD, with large-grained textured structure. As shown in Fig. 7, the photocurrent begins to saturate at about 10^5 V cm^{-1} and the best-fit of data, allowing a separate evaluation of mobility and lifetime (see Eq. (9) and (10)), gives $\mu \approx 300$ cm^2 V^{-1} s^{-1}, $\tau \approx 0.5$ ns and hence $\mu\tau \approx 10^{-7}$ cm^2 V^{-1}. As expected, the mobility value is intermediate between those of single crystal and small-grained polycrystal samples. Similarly the lifetime value is intermediate between the few nanoseconds estimated in single crystal natural diamond and 0.1 ns evaluated in small grained HFCVD diamond. It is noteworthy that comparable $\mu\tau$ values are obtained employing pulsed illumination and deep UV excitation. In particular, irradiating a thin (20 μm) textured film, grown by MW-PACVD, with 125 nm laser pulses, peak photocurrent values versus applied field exhibit the behaviour shown in Fig. 8. At variance with the measurements reported above, here light penetration depth in diamond is lower than 1 μm, so that only the response of surface layers is checked. However the evaluated $\mu\tau$ product is again in the 10^{-8}–10^{-7} cm^2 V^{-1} range.

As a conclusion, on going from single crystals to textured polycrystals to unoriented microcrystals, minority carrier $\mu\tau$ products in diamond reduce from about 10^{-5} to 10^{-7} to 10^{-8} cm^2 V^{-1}, respectively. In correspondence both mobility and lifetime separately drop, but the former decreases for about two orders of magnitude (typically from 2500 to 40 cm^2 V^{-1} s^{-1}) while the latter moves only in the range 0.1–1.0 ns.

It is noteworthy that in photocurrent versus field plots

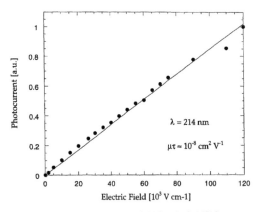

Fig. 6. Photocurrent versus applied field for ohmic MSM structure based on unoriented small-grained HFCVD diamond film. Best-fit curve according to Eq. (2), Eq. (3), Eq. (7).

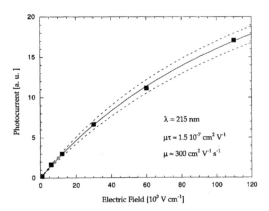

Fig. 7. Photocurrent versus applied field for ohmic MSM structure based on textured large-grain MW-PACVD diamond film. Best-fit curve (continuous line) as in Fig. 6. Theoretical predictions according to Eq. (3) and Eq. (7), allowing a ±10% variation of the fitting parameters, are also reported (dotted lines) in order to appreciate the precision of the method to obtain μ and τ values.

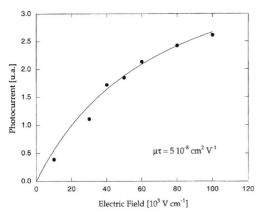

Fig. 8. Peak photocurrent values at different applied field strengths for MW-PACVD textured diamond film under 125 nm laser pulse irradiation. Best-fit curve as in Fig. 6.

the saturation field is $\mathscr{E}_{sat} = \nu_{sat}/\mu$ and hence materials with higher mobility exhibits saturation behaviour at lower field, in agreement with the experimental data.

Mobility decrease appears to be related to diamond morphology, in particular to the presence of grain boundaries and hence to grain size, whereas lifetime variation seems to be less sensitive to morphology, suggesting that recombination and/or deep trapping processes mainly involves intragrain structural defects (e.g. dislocations). Relating lifetime values to midgap state density N_{gap} by the relationship:

$$\frac{1}{\tau} = \sigma_c \nu_{th} N_{gap} \tag{12}$$

(where σ_c is the capture cross section and ν_{th} the thermal

velocity) and assuming for σ_c values in the range 10^{-16}–10^{-13} cm^2 [13], we obtain $N_{gap} \approx 10^{16}$–10^{18} cm^{-3}. Such a large midgap state density seems to be present in diamond films independently of grain dimension and orientation.

A practical consequence of this situation is shown in Fig. 9, where we report the transient photoconductivity signals measured in single crystal [14], textured [14] and randomly oriented diamond samples under simultaneous 125 and 313 nm laser pulses: in all cases the same pulse shape is practically observed, with pulse width at half maximum around 7 ns.

5. Conclusions

Steady-state measurements of photocurrent versus applied field in planar metal–diamond–metal structures are shown to allow reliable estimates of minority carrier mobility–lifetime products. Their values typically occur in the 10^{-8}–10^{-5} cm^2 V^{-1} range, depending on diamond morphology and structure. Lower $\mu\tau$ values are generally found in small-grain randomly oriented films whereas higher values refer to large-grain textured films and to single-crystal diamond. However such large differences mainly reflect variations in mobility values, while minority carrier lifetime typically ranges between 10^{-10} and 10^{-9} s and exhibits only a minor dependence on diamond structure.

Mobility variations are mainly related to the scattering efficiency of intergrain defects and impurities, while possible reasons for subnanosecond lifetimes are related to mid-gap density of states, where contributions from intragrain defects seem to be dominant and largely independent of diamond morphology.

Acknowledgements

Authors are grateful to Dr. P. Ascarelli and Dr. C. Nebel for providing HF-CVD and MW-PACVD diamond samples, and to Prof. M. C. Castex for making available UV laser set-up.

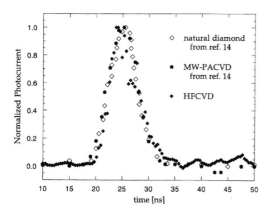

Fig. 9. Photoconductive signal measured in single crystal IIa, MW-PACVD and HFCVD diamond samples under simultaneous irradiation at 313 and 125 nm.

References

[1] Ascarelli P, Cappelli E, Mattei G, Pinzari F, Martelli S. Diamond Rel Mater 1995;4:464.
[2] Fueßer HJ, Roesler M, Hartweg M, Zachai R, Jiang X, Klages C-P. Proceedings of the Third International Symposium on Diamand Materials. The Electrochemical Society 1993;93(17):102.
[3] Museur L, Zheng WQ, Kanaev AV, Castex MC. IEEE Journal of Selected Topics in Quantum Electronics 1995;1:900.

[4] Bhattacharya P. Semiconductor optoelectronic devices. London: Prentice–Hall, 1994:chap. 8.

[5] Chu V, Conde JP, Shen DS, Wagner S. Appl Phys Lett 1989;55(3):262.

[6] Crandall RS, Balberg I. Appl Phys Lett 1991;58(3):508.

[7] Ng KK. Complete guide to semiconductor devices. New York: McGraw–Hill, 1995:548.

[8] Nava F, Canali C, Artuso M, Gatti E, Manfredi PF. IEEE Trans Nucl Sci NS– 1979;26(1):308.

[9] Binari SC, Marchywka M, Koolbeck DA, Dietrich HB, Moses D. Diamond Rel Mater 1993;2:1020.

[10] Han S, Pan LS, Kania DR. In: Pan LS, Kania DR, editors. Diamond: electronic properties and applications. USA: Kluwer Academic Publisher. 1995:chap. 6.

[11] Kania DR, Pan LS, Bell P, Landen OL, Kornblum H, Pianetta P. J Appl Phys 1990;68(1):124.

[12] Pan LS, Han S, Kania DR, Zhao S, Gan KK, et al. J Appl Phys 1993;74(2):1086.

[13] Chen W, Milnes AG. Ann Rev Mater Sci 1980;10:157.

[14] Foulon F, Bergonzo P, Borel C, Marshall RD, Jany C, Besombes L, Brambilla A, Riedel D, Museur L, Castex MC, Giacquel A. J Appl Phys 1998;84:5331.

Pergamon

CARBON

Carbon 37 (1999) 817–822

High-performance devices from surface-conducting thin-film diamond

Richard B. Jackman[a],[*], Hui Jin Looi[a], Lisa Y.S. Pang[a], Michael D. Whitfield[a], John S. Foord[b]

[a]*Electronic and Electrical Engineering, University College London, Torrington Place, London WC1E 7JE, UK*
[b]*Physical and Theoretical Chemistry, University of Oxford, South Parks Road, Oxford OX1 3QZ, UK*

Abstract

Early predictions that diamond would be a suitable material for high-performance high-power devices were not supported by the characteristics of diodes and field effect transistors (FETs) fabricated on boron doped (p-type) thin-film material. In this paper commercially accessible polycrystalline thin-film diamond has been turned p-type by the incorporation of near-surface hydrogen, a type of film often referred to as 'surface conducting'. Schottky diodes and metal-semiconductor FETs (MESFETs) have been fabricated using this approach which display unprecedented performance levels; diodes with a rectification ratio$>10^6$, leakage currents<1 nA, no indication of reverse-bias breakdown at 100 V and an ideality factor of 1.1 have been made. Simple MESFET structures that are capable of withstanding V_{DS} values of 100 V with low leakage and current saturation (pinch-off) characteristics have also been fabricated. Predictions based upon experiments performed on these devices suggest that optimised device structures will be capable of operation at power levels up to 20 W mm^{-1}, implying that thin-film diamond may after all be an interesting material for power applications. © 1999 Elsevier Science Ltd. All rights reserved.

Keywords: A. Diamond; D. Electronic properties

1. Introduction

Diamond can display extreme electronic and physical properties, such as very high carrier mobilities, saturated carrier velocities, electric field breakdown strength and thermal conductivity [1]. This has lead to speculation that it is an ideal material for the fabrication of high-power high-frequency electronic devices; 'Keyes' and 'Johnson' figures of merit are considerably higher for diamond than for Si, GaAs or SiC [2]. The emergence of commercially accessible thin-film diamond, grown by chemical vapour deposition (CVD) methods, has enabled these predictions to be experimentally tested. The fabrication of field effect transistors (FETs) has been investigated by a number of groups but to date device characteristics have been disappointing. There are two principle reasons for this. Firstly, state-of-the-art thin-film diamond is a polycrystalline material which can contain significant concentrations

of non-diamond carbon and other impurities such as nitrogen [3]. Grain boundary and impurity scattering lead to poor carrier transport characteristics. The second is the difficulty experienced with doping diamond, which is a very dense material. Boron can be incorporated and will act as an acceptor, although it displays an activation energy of 0.37 eV which leads to less than 1% of the incorporated boron being activated at room temperature [4]. FETs formed from this type of p-type material can be operated with reasonable performance characteristics at elevated temperatures (>500 K), where the level of boron activation is considerably improved, but cannot be used satisfactorily at room temperature [5,6].

An alternative approach for the formation of p-type characteristics within single crystal diamond has recently emerged [7]. This involves the incorporation of hydrogen at, or near, the surface of the film; significant near-surface hole concentrations result. Such films are often refereed to as 'surface conducting' diamond. Single crystal diamond FETs which operate successfully at room temperature have been demonstrated [7,8]; we have also fabricated the first

*Corresponding author.
E-mail address:* r.jackman@ee.ucl.ac.uk (R.B. Jackman)

0008-6223/99/$ – see front matter © 1999 Elsevier Science Ltd. All rights reserved.
PII: S0008-6223(98)00277-2

MESFET structures to be formed on polycrystalline material using this effect [9]. The origin of these carriers remains controversial with both surface band bending and the formation of shallow acceptor states being cited [10–12], although the latter appears more likely [12–14]. We have recently shown that carrier concentrations in polycrystalline thin-film diamond as high as 10^{19} cm^{-3} can be introduced in this manner. Hall carrier mobility values of ~70 cm^2/Vs can be achieved at a carrier concentration of 5×10^{17} cm^{-3}, equalling those measured in boron doped single crystal diamond and considerably better than the 1–5 cm^2/Vs typical of boron doped polycrystalline material [15]. This paper reports upon the use of this form of surface-conducting polycrystalline thin-film diamond for the fabrication of high-performance metal-semiconductor FETs (MESFETs), compares them with boron doped diamond FETs and considers their potential use within power applications.

2. Experimental methods

Free standing (~300 μm thick) polycrystalline diamond, grown by microwave plasma enhanced CVD, was used throughout. The film comprised randomly aligned grains, 20–50 μm in size, as shown in Fig. 1. No purposeful doping was carried out and the growth chamber had not been previously used for boron doping experiments. Raman analysis (Renishaw system 2000 with (red) He–Ne laser light) revealed only the presence of a sharp peak at 1332 cm^{-1}, indicative of good quality diamond [1]. Secondary ion mass spectrometry (SIMS) showed the presence of a 20 nm hydrogen rich region within the top surface of the as-grown material; Hall effect measurements in similar material indicated a carrier concentration of ~10^{19} holes cm^{-3}. Heating for periods greater than 1 h at 200°C (in air or vacuum) reduced this to a value of 5×10^{17}

cm^{-3}, a value that was not changed by further heat treatments up to this temperature. Metal contacts were formed by thermal evaporation; simple shadow masking was employed to form rectangular stripes with a dimension of 3.5×0.7 mm and a separation of 0.5 mm to assess the I–V characteristics of both Al and Au. Simple MESFET structures were formed using an Al gate (thickness 200nm) and point ohmic contacts (gold coated probes, estimated contact area and hence gate width, 80 μm) 200 μm from each side of the gate. Current–voltage (I–V) measurements were performed using a Hewlett Packard semiconductor parameter analyser (4145B).

3. Results

Metal contacts formed on 'as-grown' surfaces revealed near ideal characteristics; Al formed a Schottky contact with a measured barrier height of 0.98 eV whilst Au contacts were near-ohmic in nature. This behaviour can be understood in terms of the differing electronegativities of the metals and is indicative of a diamond surface which displays little Fermi level pinning. Schottky barrier diodes with greater than six orders of magnitude rectification ratio and an ideality factor of 1.1 were routinely fabricated; an example is shown in Fig. 2a. A reverse-bias leakage current of less than 1 nA is apparent with no indication of reverse-bias breakdown, even at 100 V. It is important to note that at room temperature the forward current level for this device is in the mA range. Other devices were fabricated following heating of the 'as-grown' diamond film to 200°C for 4 h. The I–V characteristics achieved at room temperature (Fig. 2b) indicate that a good Schottky barrier diode is still formed, with low reverse-bias leakage current, although it is now less conductive when in its forward-bias condition. Fig. 2b also shows the I–V characteristics measured at elevated temperatures for these

Fig. 1. Optical micrograph image of the polycrystalline diamond films used throughout the present study.

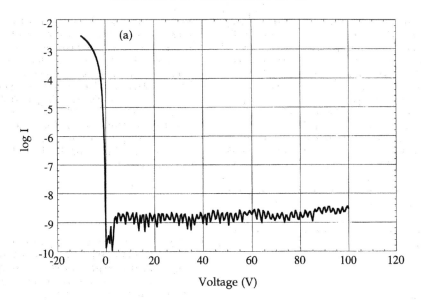

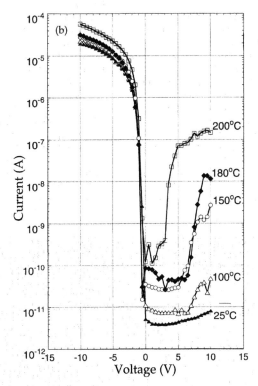

Fig. 2. (a) *I–V* characteristics for Al–Au contacts at room temperature. (b) *I–V* plot of Al–Au contacts following heat treatment at temperatures up to 200°C.

devices; reverse-bias leakage current levels increase (to around 0.1 μA at 200°C), but only a modest increase in forward-bias current levels is apparent.

Simple MESFET structures were fabricated with Au source and drain point contacts and an Al gate. With the gate unbiased, little drain–source current (I_{DS}) could be measured. For gate bias levels greater than −1.0 V, I_{DS} increased with increasing drain–source voltage (V_{DS}), indicative of an enhancement mode MESFET structure. Fig. 3a shows the variation of I_{DS} with V_{DS} as a function of applied gate bias; gate promoted channel 'pinch-off' is clearly occurring leading to saturation in the channel current. The gate leakage current was around 10% of the I_{DS} value. Fig. 3b shows V_{GS} plotted against the saturated value of I_{DS}; the linear relationship that is evident following 'turn-on' of the device (at $V_{GS} \sim 1$ V) suggests that these devices would be capable of modulating significantly higher channel currents than those investigated here.

4. Discussion

The near-ideal barrier heights measured here for Au and Al contacts on the 'as-grown' and heated diamond surfaces contrast sharply with those observed when boron doped diamond is used. For example, on boron doped material most metals give rise to a significant Schottky barrier whose height is more dependent upon the recent history of the diamond surface than the choice of metal [16]; the formation of ohmic contacts usually requires the formation of a reacted interface, typically a titanium carbide layer [17]. This indicates the extent to which the surface of all the diamond studied here remains hydrogenated, as hydrogen is known to readily attach itself to surface dangling bonds enabling metal–semiconductor interfaces to be formed which are relatively free from Fermi level pinning [7,9,13]. The generation of free carriers from hydrogen also leads to a significant reduction in carrier scattering (as witnessed by the high carrier mobility values recorded for this material). These effects have combined to enable us to fabricate polycrystalline diamond Schottky diodes with unprecedented *I–V* performance characteristics. An ideality factor of 1.1, allied to the absence of any breakdown at reverse-bias levels exceeding 100 V, leakage currents on the nA scale and mA forward-bias current levels for the simple structures made here (Fig. 2a) indicates the promise of this approach for the formation of high-power diodes. These characteristics contrast sharply with Schottky diodes fabricated on boron doped polycrystalline diamond which typically suffer from 'soft' breakdown at modest voltages and low forward-bias current levels [18]. Similar characteristics persist following heating to 200°C (Fig. 2b), suggesting that these devices will offer stable operation at room temperature for long periods. The fact that a previously heated device shows only a slight increase in forward-bias current level when operated at elevated temperatures (Fig. 3) suggests that the process for generating free carriers by hydrogenating diamond has a relatively low activation energy.

We have been able to produce polycrystalline diamond MESFET structures which display excellent gate control; again this has not been previously demonstrated using boron doped material. The 'normally off' enhancement mode characteristics are consistent with our SIMS estimate that the carriers are distributed over a region less than 20 nm thick; such a layer is thin enough to be fully depleted by the gate contact under zero bias. The devices are readily driven into saturation and capable of switching potentials greater than 100 V (Fig. 3). The relatively high gate leakage current (~10% of I_{DS}) can be understood in terms of the turn-on potential of the device which is similar to the Schottky barrier height of the Al gate. This means that one of the gate-drain contact will effectively be forward biased during operation leading to the gate leakage observed. This could be prevented in more carefully designed device structures by modifying, for example, the depth of the doped region.

The potential power handling capability of these simple devices can be estimated by considering the effect of the source and drain point contacts used here. Evaporated contacts with a length more closely matched to the full gate width available, displayed a decrease in series resistance of around 25 times. On this basis the devices, had they used proper contact pads, would have a power handling capability of around 0.2 W mm^{-1} at a gate bias of −5.5V, with a V_{DS} value of 100 V; the normalised transconductance is estimated to be 0.47 mS mm^{-1} at room temperature, which is the highest yet reported value for polycrystalline diamond devices. Indeed, no examples of room temperature operation of other forms of polycrystalline diamond MESFETs can be found in the literature. MISFET devices, using an insulating diamond gate and a boron doped channel, can display transconductance values of around 5 μS mm^{-1} [5]. The characteristics achieved during the current study arise despite the source and drain being separated by 0.5 mm, compared to the micron scale encountered in state-of-the-art power devices. To enable a comparison to be made with such devices, the effect of the additional series resistance introduced between the source and drain must be considered. We have previously carried out Hall effect measurements on similar material to that used in the present study, to obtain sheet resistivity values, and compared these to *I–V* characteristics measured by the transfer-length method for deconvoluting specific contact resistance and sheet resistivity [19]. From these values it can be estimated that the present devices would be one order of magnitude more conductive if the source to drain separation was around 5 μm; this would lead to a power handling capability at a V_{DS} value of 100 V of 2 W mm^{-1}, which compares favourably with

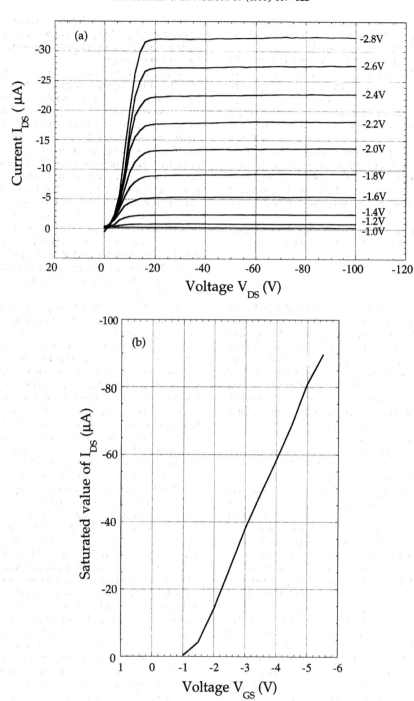

Fig. 3. (a) MESFET source to drain *I–V* characteristic plotted as a function of increasing gate bias. (b) MESFET source to drain current measured at $V_{DS} = 100$ V for increasing gate bias.

state-of-the-art SiC devices [20]. This value would be significantly greater if the specific contact resistance of the Au contacts did not become dominant within structures with reduced dimensions. It has been shown that Si-doped Al contacts on p-type diamond can display specific contact resistance values of ~10^{-7} $\Omega\,cm^2$ [21], compared to values of around 10^{-3} $\Omega\,cm^2$ for the Au used here. The conductivity of the devices could therefore be further increased by optimising the ohmic contacts leading to a power handling capability somewhere between the 2 W mm^{-1} predicted for devices using Au contacts and 20 W mm^{-1} that would arise if the specific contact resistance was negligible.

5. Concluding remarks

Thin-film polycrystalline diamond Schottky diodes and MESFETs have been produced which display unprecedented performance levels. The primary origin of the improvement achieved is the use of near-surface hydrogen to produce p-type diamond as opposed to the incorporation of boron. The process(es) which enable hydrogen to create p-type characteristics appear to be almost fully activated at room temperature and device structures with stable characteristics can be fabricated and operated at temperatures as high as 200°C. This has enabled us to make realistic predictions for the power handling capabilities of this form of device. Modest scaling of the simple devices fabricated here would lead to a MESFET structure capable of operation at 0.2 W mm^{-1} (with a V_{DS} value of 100 V); optimised structures may operate at powers as high as 20 W mm^{-1}, which is considerably greater than, for example, state-of-the-art SiC devices. When other properties of diamond are considered (thermal conductivity, electric field breakdown strength, radiation hardness etc . . .) it would appear that commercially accessible polycrystalline thin-film diamond based devices may be suitable for several power device applications.

References

[1] Geis MW, Rothman DD, Ehrlich DJ, Murphy RA, Lindley WT. IEEE Electron Dev Lett 1987;8:341.

[2] Gildenblat GSh, Grot SA, Badzian A. Proc of IEEE 1991;79:647.

[3] Sonoda S, Won JH, Yagi H, Hatta A, Ito T, Hiraki A. Appl Phys Lett 1997;70:2574.

[4] Werner M, Dorsch O, Baerwind HU, Obermeier E, Haase L, Seifert W, Ringhandt A, Johnston C, Romani S, Bishop H, Chalker PR. Appl Phys Lett 1994;64:595.

[5] Nishimura K, Kumagai K, Nakamura R, Kobashi K. J Appl Phys 1994;76:8142.

[6] Pang LYS, Chan SSM, Jackman RB. Appl Phys Lett 1997;70:339.

[7] Kawarada H, Aoki M, Ito M. Appl Phys Lett 1994;65:1563.

[8] Hokazono A, Ishikura T, Nakamura K, Yamashita S, Kawarada H. Diamond and Relat Mater 1996;5:706.

[9] Looi HJ, Pang LYS, Wang Y, Whitfield MD, Jackman RB. IEEE Elect Dev Letts 1998;19:112.

[10] Shirafuji J, Sugino T. Diamond and Relat Mater 1995;5:706.

[11] Kawarada H, Sasaki H, Sato A. Physics Rev B 1995;52:11351.

[12] Hayashi K, Yamanaka S, Okushi H, Kajimura K. Appl Phys Lett 1996;68:376.

[13] Looi HJ, Pang LYS, Molloy AB, Jones F, Foord JS, Jackman RB. Diamond and Relat Mater 1998;7:550.

[14] Looi HJ, Foord JS, Jackman RB. Appl Phys Lett 1998;72:353.

[15] Malta DM, Windheim JA, Fox BA. Appl Phys Lett 1993;62:2926.

[16] Mori Y, Kawarada H, Hiraki A. Appl Phys Lett 1990;58:940.

[17] Shiomi H, Nakahata H, Imai T, Nishibayashi Y, Fujimori N. Jpn J Appl Phys 1989;28:758.

[18] Gildenblat GSh, Grot SA, Hatfield CW, Badzian AR, Badzian T. IEEE Electron Dev Lett 1993;11:371.

[19] Looi HJ, Foord JS, Jackman RB, J Appl Phys, in press.

[20] Sriram S, Siergiej RR, Clarke RC, Agarwal AK, Brandt CD. Physica Status Solidi A: Applied Research 1997;162:441.

[21] Werner M, Dorsch O, Baerwind HU, Obermeier E, Johnston C, Chalker PR, Romani S. IEEE Trans on Electron Devices 1995;42:1344.

Pergamon

Carbon 37 (1999) 823–827

CARBON

Electron energy-loss spectroscopy in transmission of undoped and doped diamond films

S. Waidmann[a,*], K. Bartsch[a], I. Endler[a], F. Fontaine[b], B. Arnold[a], M. Knupfer[a], A. Leonhardt[a], J. Fink[a]

[a]*Institut für Festkörper- und Werkstofforschung Dresden, Postfach 270016, D-01171 Dresden, Germany*
[b]*Forschungszentrum Rossendorf, Institut für Ionenstrahlphysik und Materialforschung, Postfach 510119, D-01314 Dresden, Germany*

Received 17 June 1998; accepted 3 October 1998

Abstract

We have prepared thin diamond films by microwave assisted plasma chemical vapour deposition (MWCVD) and hot filament chemical vapour deposition (HFCVD). Diamond powder pre-treatment of the silicon substrates or bias potential were used for nucleation enhancement. Doping of the films was carried out in two ways: in-situ during the CVD process or after the deposition by ion implantation. Scanning electron microscopy (SEM) and X-ray diffraction (XRD) have been applied to characterize the morphology and texture. Electron energy-loss spectroscopy in transmission was then used to investigate the electronic structure of the diamond films as a function of the preparation parameters and the doping level. © 1999 Elsevier Science Ltd. All rights reserved.

Keywords: A. Diamond; C. Electron energy-loss spectroscopy (EELS)

1. Introduction

Due to its impressive thermal and electronic properties, there is considerable interest in the use of diamond for electronic applications. Especially in a harsh and hot environment, diamond could be the material of choice. Natural diamond and synthetic high pressure diamond are both too expensive for their mass application in electronics, but chemical vapour deposition represents an economically feasible method to grow high quality diamond thin films for technical applications. The growth of diamond on silicon substrates is of special interest because single crystalline silicon substrates are widely available with very high surface quality. Large area single crystalline growth on silicon has not been achieved up to now but nevertheless local heteroepitaxial growth with almost perfectly oriented diamond grains has been achieved [1].

While *p*-type and *n*-type doping of silicon is state-of-the-art, doping of diamond still remains unsatisfactory. The

obvious dopants in diamond which are also found as natural impurities are nitrogen and boron. Isolated substitutional nitrogen atoms in diamond are deep donors with a donor level at 1.7 eV below the top of the conduction band and thus cannot be used for electronic applications. Substitutional boron is a relatively shallow acceptor with an energy level at 0.37 eV above the top of the valence band [2]. Consequently, boron doping has been widely studied in previous years but there remain open questions which render further work necessary [3]. The most promising methods to dope diamond are in-situ doping during the growth process [4] and ion implantation [5].

We have prepared diamond films using microwave assisted plasma chemical vapour deposition (MWCVD) and hot filament chemical vapour deposition (HFCVD). In-situ doping with trimethylborate (TMB) has been carried out in order to deposit boron-doped diamond films. The results are compared with those from diamond films which were boron ion implanted. The film morphology was investigated using scanning electron microscopy (SEM). X-ray diffraction (XRD) was applied to characterize their texture and crystallinity and electron energy-loss spectroscopy (EELS) in transmission was performed to

---

*Corresponding author.

Table 1
Deposition parameters of the diamond samples prepared using MWCVD and HFCVD

Sample	MW1	MW2	MW3	HF1
Microwave power (W)	310	475	475	–
Substrate temperature (°C)	790	845	845	900
Pressure (mbar)	40	40	40	40
CH_4/H_2 (at.%)	0.4	1.2	1.2	0.9
O_2/H_2 (at.%)	0.25	–	–	–
Trimethylborate (at. ppm)	–	–	104[a]	–
Thickness of doped surface layer (μm)	–	–	2	–
Total film thickness (μm)	5.2	7.5	9.5	$\approx$8

[a]Only the surface layer has been doped.

obtain detailed information regarding their electronic structure.

2. Experimental

Diamond film deposition was performed in a microwave plasma assisted chemical vapour deposition (MWCVD) reactor and a hot filament chemical vapour deposition (HFCVD) reactor, respectively. As substrates we used polished Si (100) wafers. Two different methods were applied to enhance the nucleation density: for the HFCVD experiments we used diamond powder pretreatment and for the MWCVD experiments bias enhanced nucleation. It has been shown that heteroepitaxial growth of diamond on β-SiC and Si can be achieved by the use of bias enhanced nucleation [6,7].

The deposition parameters of the prepared diamond films are given in Table 1. The samples MW1–3 were prepared using MWCVD, the sample MW3 was prepared with the same deposition parameters as MW2 except the doping step. In the case of MW3 a surface layer of ~2 μm thickness was boron-doped via the addition of 100 ppm trimethylborate (TMB) to the gas flow in the reactor. According to Ref. [4] this concentration should approximately lead to a boron concentration in the range of 0.1 to 0.2 at.%.

The sample HF1 was prepared with HFCVD. One half of the as-grown sample was boron ion implanted resulting in a boron concentration of about 0.28 at.%. The implanta-

tion parameters are given in Table 2. The implantation profile was calculated by the Monte-Carlo simulation program TRIM. In order to minimize graphitization, implantation was carried out at 1000°C. It has been demonstrated that high temperature implantation dynamically reduces the formation of defects during the implantation process [8]. No additional annealing step after implantation was carried out.

EELS in transmission requires free standing thin films with a thickness not exceeding 100 nm. Therefore, the first step of the film preparation for EELS was to remove the silicon substrate with potassium hydroxide. Subsequently, the diamond film itself was thinned using ion sputtering (5 keV Ar^+ ions) to obtain a 100-nm thin film from the surface region of the original film. The EELS measurements were performed in transmission with a purpose-built spectrometer (primary energy: 170 keV) which is described in detail in Ref. [9]. An energy resolution of 160 meV and a momentum resolution of 0.06 Å^{-1} were chosen.

3. Results and discussion

3.1. SEM and XRD

The morphology and texture of the CVD-grown diamond samples were characterized with SEM and XRD. Figure 1 shows SEM images and XRD profiles of the diamond films which were prepared using MWCVD. The sample MW1 presented in Fig. 1a shows a rough $\langle 110 \rangle$

Table 2
Process parameters of the ion implantation procedure

Sample	Ion	Energy (keV)	Dose ($10^{15}/cm^2$)	Max. implantation depth (μm)	Concentration (at.%)
HF1	B	150	4.6		
	B	118	3.07		
	B	93	2.75		
	B	71	2.62		
	B	52	2.36		
	B	36	4.0		
				0.2	0.28 (total)

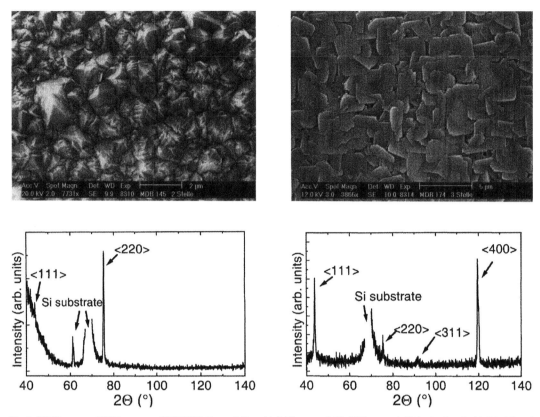

Fig. 1. SEM images and XRD patterns of MWCVD diamond films: (a) ⟨110⟩ textured; (b) ⟨100⟩ textured; (c) in-situ doped with 0.1–0.2 at.% boron.

morphology with etching grooves due to the use of additional oxygen during growth and an XRD pattern which indicates a pronounced ⟨110⟩ texture of the film. The SEM image of sample MW2 (Fig. 1b) exhibits a ⟨100⟩ morphology which is partly biaxial and an XRD spectrum showing the clear ⟨100⟩ texture. In Fig. 1c the SEM image and XRD pattern of the sample MW3, which had been doped in-situ with boron, are presented. As mentioned above, the only difference between MW2 and MW3 is the boron-doped surface layer of 2 μm thickness. One can see that the addition of trimethylborate (TMB) of 100 ppm during the MWCVD growth destroys the ⟨100⟩ morphology in agreement with Ref. [4]. The corresponding XRD spectrum shows an analogous behaviour, as the ⟨100⟩ texture has been diminished in the doped film.

We note that according to Ref. [4] only the morphology and texture of the ⟨100⟩ textured films are altered to an irregular form as a consequence of TMB addition, whereas ⟨110⟩/⟨111⟩ textures remain unchanged. This might be due to the oxygen present in TMB. Previous work on non-

oxygen-containing boron precursors has not shown a deterioration of the ⟨100⟩ morphology in connection with boron doping [10]. Instead, boron in low concentrations can even enhance the smoothness of ⟨100⟩ surfaces [1].

3.2. EELS

Figure 2 shows C 1s excitation edges of undoped and doped diamond samples. Such measurements probe excitations of the 1s electrons into the unoccupied 2p-derived states. The covalent bonds between the sp^3 hybridized carbon atoms in perfect diamond are of pure σ-character. The onset of the C 1s excitation into the unoccupied σ*-states lies at 289.1 eV and forms the well-known core exciton which is visible as a sharp peak. In contrast, sp^2 hybridized systems, such as graphite or amorphous carbon, have additional unoccupied π*-states, with a corresponding C 1s excitation onset around 285 eV. Thus, C 1s excitation studies provide a powerful tool to unambiguously distinguish between diamond on the one hand and

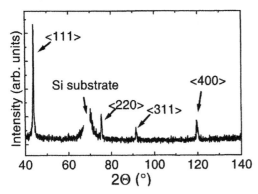

Fig. 1. (*continued*)

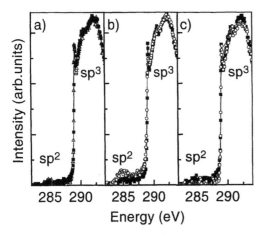

Fig. 2. C 1s excitation spectra of: (a) ⟨110⟩ textured MWCVD diamond (squares) and natural diamond type IIa (open triangles); (b) undoped (squares) and ion implanted diamond (0.28 at.% boron, open circles); (c) undoped (squares) and in-situ doped diamond (0.1–0.2 at.% boron, open circles).

graphite or amorphous carbon on the other hand. The ratio of the integrated intensities of the π^*-states to the σ^*-states in comparison to graphite gives a quantitative measure of the carbon atoms with sp^2 hybridization present in the sample under consideration [11,12].

A comparison of the C 1s excitation edge of natural diamond type IIa with the ⟨110⟩ textured MWCVD diamond film MW1 (Fig. 2a) shows a similarly low amount of unoccupied sp^2 states in both samples. This demonstrates the high purity of the surface region of the ⟨110⟩ textured diamond film grown in this study which can be attributed to the low methane content and the additional use of oxygen in the process gas. The very small sp^2 amount that is visible even in the spectrum of the type IIa diamond can be traced back to an amorphous surface layer caused by the ion-beam thinning [13].

Figure 2b shows the C 1s excitation spectra of the undoped and boron-implantation-doped sample HF1. One can clearly see that the sp^2 amount after ion implantation is more than doubled [14]. It is well known that ion implantation can generate damaged graphite-like regions in diamond [8]. Although high temperature implantation and boron doses low enough to minimize graphitization were chosen, it is clear from Fig. 2b that this could not be avoided completely. This situation might be improved by further optimization of the implantation parameters (higher implantation temperature and/or post anneal) or by additionally removing the uppermost surface layer after implantation as this layer is most strongly affected by the implantation process.

In contrast to this, one can see in Fig. 2c that in-situ boron doping of microwave assisted CVD films to an estimated concentration of 0.1–0.2 at.% even reduces the sp^2 amount in comparison to the undoped sample. A similar result has also been found previously [13] and it has been proposed [15] that p-type doping enhances the vacancy diffusion and therefore leads to a better annealing of defects. This might explain the result shown in Fig. 2c. As far as the amount of sp^2 carbon in the doped sample is concerned, one could speculate that in-situ boron doping of diamond is the more suitable preparation method. However, we note that the sp^2 content of the samples is only one of several parameters which determine their technical applicability.

4. Conclusions

We have shown that high resolution EELS in transmission is an excellent method to determine the quantity of sp^2 carbon in diamond, which is attributed to graphitic or amorphous carbon phases. Boron doping by ion implantation (for the implantation parameters used here) enhances the sp^2 content, which can be explained by ion-induced damage upon implantation. In contrast, in-situ doping up to a concentration of 0.1–0.2 at.% boron reduces the sp^2

amount in comparison to the undoped case. A possible explanation of this effect could be an enhanced diffusion of vacancies which then leads to a lower defect concentration.

Acknowledgements

This work was supported financially by the Sächsische Staatsmisterium für Wissenschaft und Kultur. A. Zetzsche is thanked for help in carrying out the MWCVD experiments. We are grateful to S. Pichl for taking the SEM images and to M.S. Golden for a critical reading of the manuscript.

References

[1] Jiang X, Schiffmann K, Klages C-P, Wittorf D, Jia CL, Urban K, Jäger W. J Appl Phys 1998;83:2511.

[2] Kajihara SA, Antonelli A, Bernholc J, Car R. Phys Rev Lett 1991;66:2010.

[3] Pang LYS, Chan SSM, Jackman RB, Johnston C, Chalker PR. Appl Phys Lett 1997;70:339.

[4] Locher R, Wagner J, Fuchs F, Maier M, Gonon P, Koidl P. Diamond Relat Mater 1995;4:678.

[5] Prawer S, Uzan-Saguy C, Braunstein G, Kalish R. Appl Phys Lett 1993;63:2502.

[6] Stoner BR, Glass JT. Appl Phys Lett 1992;60:698.

[7] Jiang X, Klages C-P. Diamond Relat Mater 1993;2:1112.

[8] Kalish R, Prawer S. Nucl Instrum Methods Phys Res 1995;B106:492.

[9] Fink J. Adv Electron Electron Phys 1989;75:121.

[10] Liu K, Zhang B, Wan M, Chu JH, Johnston C, Roth S. Appl Phys Lett 1997;70:2891.

[11] Egerton RF. Electron energy-loss spectroscopy in the electron microscope, 2nd ed. New York: Plenum Press, 1996.

[12] Berger SD, McKenzie DR, Martin PJ. Phil Mag Lett 1988;57:285.

[13] Polo MC, Cifre J, Esteve J. Vacuum 1994;45:1013.

[14] The σ^* core exciton in the spectrum of the ion-doped film (Fig. 2b) is reduced in intensity in comparison to the undoped case. This results from the use of a lower energy resolution ($\Delta E = 300$ meV, compared to 160 meV) for the measurement of the doped sample.

[15] Bernholc J, Antonelli A, Del Sole TM, Bar-Yam Y, Pantelides ST. Phys Rev Lett 1988;61:2689.

Pergamon

Carbon 37 (1999) 829–833

CARBON

Photoconductivity and recombination in diamond-like carbon

A. Ilie, J. Robertson*, N. Conway, B. Kleinsorge, W.I. Milne

Department of Engineering, Cambridge University, Cambridge CB2 1PZ, UK

Received 16 June 1998; accepted 3 October 1998

Abstract

The steady-state photoconductivity of tetrahedral amorphous carbon (ta-C) and hydrogenated ta-C (ta-C:H) has been studied as a function of temperature, light intensity, and photon energy, in order to understand the recombination process in diamond-like carbon. It is found that the levels demarking the recombination states can span only part of the gap, so that the recombination centres can vary from every defect, to some defects, to some tail states, according to conditions. © 1999 Elsevier Science Ltd. All rights reserved.

Keywords: A. Diamond-like carbon; D. Photoconductivity

1. Introduction

Diamond-like carbon (DLC) is a semiconducting form of amorphous carbon (a-C) with a significant fraction of sp3 bonding [1–3]. Types of DLC with high sp3 content are called tetrahedral amorphous carbon (ta-C) or ta-C:H and can be made by special, ion dominated deposition processes [4,5]. The electronic structure of DLC differs from that of for example a-Si:H because of the behaviour of the sp2 sites. The sp2 sites possess π states and these states form the valence and conduction band edges. The sp2 sites also tend to cluster, so that the band edges are subject to very large antisymmetric fluctuations, with a narrow gap in sp2 regions and a wide gap in sp3 regions [6,7]. DLCs are also more disordered than a-Si:H, with Urbach slopes of 0.15–0.4 eV and paramagnetic defect densities of 10^{18}–10^{20} cm^{-3}. This has a strong effect on the electronic properties. a-C:H can have strong, room temperature luminescence (PL).

The nature of the recombination centres which quench the PL has been debated [6–11]. There are strong correlations of PL efficiency with paramagnetic defect density [6,7,10], but Schutte et al [9] gave a critical case where efficiency did not correlate with defect density, and recently Giorgis et al. [11] proposed that all states within a certain energy range act as recombination centres.

*Corresponding author.

Photoconductivity is another method to probe recombination processes. DLCs have relatively low mobility so the photoconductivity is not so high, but nevertheless the results are able to resolve many questions concerning the nature of the recombination [12].

2. Experimental method

The ta-C films were deposited at room temperature using a filtered cathodic vacuum arc (FCVA) system from 99.999% pure graphite. The films were grown on Corning 7059 glass, and an 13.6 MHz RF bias was applied to the substrate to bring the incident ion energy to 100 eV, in order to maximise the sp3 content of the ta-C [13]. The ta-C:H films were grown in a low pressure Plasma beam source (PBS) using acetylene or methane as source gases [14]. Again, to maximise sp3 content, the films were deposited at an ion energy of 200 eV for acetylene or 130 eV for methane. The sp3 contents were determined by electron energy loss spectroscopy (EELS) to be 80% for ta-C and about 70% for ta-C:H, and the H content to be about 30% by ERDA. The defect densities were measured by electron spin resonance (ESR).

The photoconductivity measurements were carried out on 40–100 nm thick films using single gap or interdigitated array cells, with thermally evaporated Al and an

0008-6223/99/$ – see front matter © 1999 Elsevier Science Ltd. All rights reserved.
PII: S0008-6223(98)00279-6

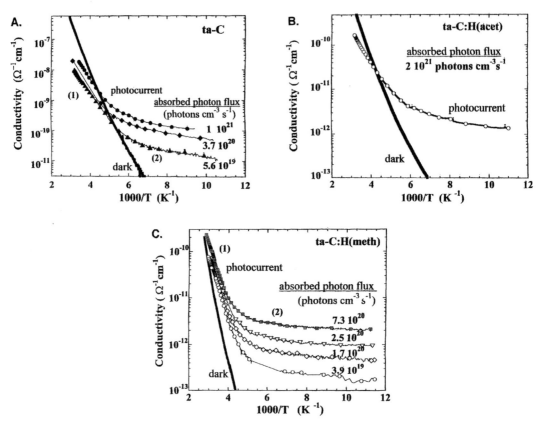

Fig. 1. Dark and photoconductivity of (a) ta-C, (b) ta-C:H from acetylene, and (c) ta-C:H from methane.

initial layer of Cr as electrodes. The heating and cooling rates were slow, to avoid thermally stimulated currents.

3. Results

Figure 1(a–c) show the variation of the dark and white-light photoconductivity of ta-C, ta-C:H(methane) and ta-C:H(acetylene) as a function of temperature. The absorbed photon fluxes are derived from the known optical absorption spectra. The photoconductivity is seen to be thermally activated and similar to the dark conductivity for tempera-

tures above ~200 K. Below 200 K, the photoconductivity varies slowly with temperature, and for ta-C:H(meth) becomes almost independent of temperature. The defect densities of each film are given in Table 1.

The photoconductivity often varies with light intensity or generation rate G according to the power law, $\sigma = kG^{\gamma}$. Figure 2 shows for white light excitation that γ varies from under 0.5 at room temperature towards 1 at low temperature.

The optical absorption edge of ta-C and ta-C:H is quite broad. Figure 3 shows the variation of photoconductivity of ta-C:H(meth) with monochromatic excitation.

Table 1

	ta-C	ta-C:H(acet)	ta-C:H(meth)
E_{04} (eV)	2.5	1.9	2.5
spin density (cm^{-3})	5×10^{20}	10^{20}	5×10^{19}
σ_d (300K)	8.3×10^{-8}	3.8×10^{-11}	1×10^{-11}
σ_{pc} (300K)	2×10^{-8}	1.2×10^{-11}	6.2×10^{-11}
σ_{pc}/σ_d (300 K)	0.2	0.32	6.6

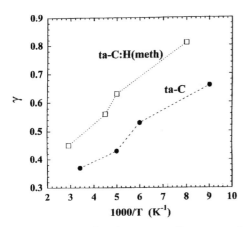

Fig. 2. Variation of the intensity power γ with temperature for white light excitation.

4. Discussion

Ta-C and ta-C:H are known to be p-type conductors [13,14]. The photoconductivity can be expressed as

$$\sigma = e\mu p = eG\mu\tau$$

where G is the photogeneration rate, μ is the hole mobility and τ is the hole lifetime. The $\mu\tau$ product can be related to the density of recombination centres N and their capture cross-section s by [15]

$$\mu\tau N = \frac{e}{4\pi kTd} = \frac{e}{4\pi kTs^{1/2}}$$

where we have assumed that transport is diffusive such that the lifetime is given by

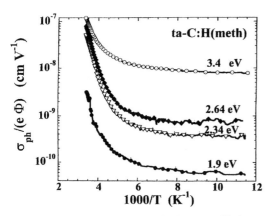

Fig. 3. Variation of photoconductivity of ta-CH(meth) with photon excitation energy.

$$1/\tau = 4\pi dDN,$$

and D is carrier diffusion coefficient and $s = \pi d^2/4$.

Thus, the $\mu\tau$ product should vary inversely with the density of recombination centres. The dangling bond defects are the recombination centres in a-Si:H, and it is natural to expect them to act in a similar manner in ta-C and ta-C:H. Comparison of Fig. 1 and Table 1 shows that the photoconductivity of ta-C is greater than that of ta-C:H(meth), whereas its spin density is also greater, so that the inverse relationship does not hold for defects. Nevertheless, there is inverse correlation between the photo-sensitivity, the ratio of photo- to dark-conductivity, and the defect density. Thus photoconductivity does not fully support the idea that defects are the main recombination centre.

The temperature dependence of dark and photoconductivity to find the transport level E_t and the hole and electron quasi Fermi levels. The dark conductivity is thermally activated with a small prefactor, which indicates that conduction occurs by hopping in localized states in the valence band. The dark Fermi level E_{f0} is some 0.3–0.4 eV higher than the transport level E_t. The hole quasi Fermi level E_{fp} lies close to E_{f0} above 200 K and then moves towards E_t for $T < 200$ K, where the photoconductivity becomes flat.

For ta-C:H, the photoconductivity is larger than the dark conductivity, so that the photogenerated minority carrier (electron) density is larger than its dark value, and E_{fn} lies in the conduction tail. However, for ta-C above 200 K, the photoconductivity drops below the dark conductivity, so that in this case E_{fn} still lies in the valence tail near E_{f0}, as shown in Fig. 4.

For localized states, there exists two demarkation energies between those states which act as traps and those which act as recombination centres. A recombination centre must be able to trap both electrons and holes. The demarkation energy for electrons D_n is defined as the energy at which the emission of electrons to the conduction edge E_c, $\nu_0\exp(-[E_c - D_n]/kT)$, equals the rate of trapping of holes, $p.\nu_{th}.s$, where p is the hole concentration, ν_{th} is the hole thermal velocity and s is the trapping cross section of the states at D_n. This gives

$$E_c - D_n = kT \log\left(\frac{\nu_0}{p\nu_{th}s}\right)$$

with a similar relationship for D_p.

We see that for ta-C:H, D_p and D_n lie in their respective band tails, and span the gap and the defect states which will lie in midgap. However, for ta-C, in which the electron concentration, n, remains low, D_p will remain in the conduction tail – in fact both demarkation levels lie in the conduction band tail. Thus, in one case, most states lying in midgap are able to act as recombination centres, but when n is low both demarkation levels will lie in the

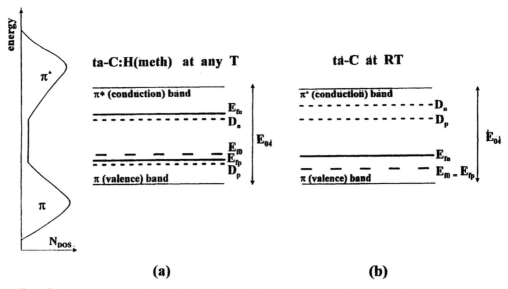

Fig. 4. Schematic of Fermi levels, transport level, and demarkation levels in (a) ta-C:H, and (b) ta-C at room temperature.

conduction band tail and now *tail states* will act as recombination centres.

We believe that the differing location of the demarkation energies is the reason why the $\mu\tau$ values do not scale inversely as the defect density. The continuing decrease of the photoconductivity seen at low temperatures in Fig. 1 shows that the demarkation energies have not yet reached a stable position in the band tails.

The second problem for identifying recombination centres is the high localization of defect states in a-C. A defect state can be occupied by 0,1 or 2 electrons as D^+, D^0 or D^-. The energy required to place the second electron on the defect level is the correlation energy, U. This is much larger in a-C:H than in a-Si:H because of the smaller Bohr radius of localized states in a-C:H. Ristein et al [16] estimated that $U \approx 3$ eV, from the ratio of the density of states to the spin density. This value of U is larger than the gap, so that most defect levels will not show both charge states in the gap, unlike a-Si:H where the $+/0$ and $0/-$ states lie within 0.3 eV of each other near midgap. Recombination requires either the successive trapping of either an electron or hole,

$$D^0 + h \rightarrow D^+ \text{ and } D^+ + e \rightarrow D^0$$

or

$$D^0 + e \rightarrow D^- \text{ and } D^- + h \rightarrow D^0$$

For hole conductors like ta-C(:H), the first reaction is the more probable step; in addition the second reaction is less likely because D^- will lie too high.

The nature of the recombination centre during PL has

been debated for some time. PL is a good test of the nature of the recombination centre because the PL efficiency should vary with the density of recombination centres N, as [15]

$$\eta = \exp\left(-\frac{4}{3}\pi R_c^3 N\right)$$

where R_c is the capture radius of the centre. This is a strong function of N, in a certain range.

Liedtke et al. [8] found that in a-SiC alloys, the PL efficiency scaled inversely with the defect density in the Si rich alloys, but not in the C-rich alloys. Schutte et al [9] then compared PL in a series of a-C:H samples prepared by PECVD on the cathode or anode with very different defect densities, and found that the PL efficiency varied strongly with band gap, but not with defect density. Robertson [6,7] argued that defects were still likely to be the dominant recombination centre as a-C:H had more similarities to a-Si:H than other semiconductor systems, and that PL still occurred in a-C:H at much higher defect densities of up to 10^{20} cm^{-3} because the Bohr radius and R_c were smaller than in a-Si:H. The problem is that the defect density in a-C:H correlates quite strongly with the band gap, so that the precise variations can be unclear. Recently, Giorgis et al [14] used samples with low bandgap and low paramagnetic defect density to show that the PL efficiency correlates best with the total density of states within some energy range near midgap, which can include the band tails in the case of narrower gap forms of a-C:H.

The analysis above shows that this is possible; both midgap states and tail states can under appropriate conditions, lie between the demarkation levels, and act as recombination centres.

References

[1] Robertson J. Adv Phys 1986;27:361.

[2] Robertson J, O'Reilly EP. Phys Rev B 1987;35:2946.

[3] Robertson J, O'Reilly EP. Surface Coatings Technol 1992;50:185.

[4] Fallon PJ, Veerasamy VS, Davis CA, Robertson J, Amaratunga G, et al. Phys Rev B 1993;48:4777.

[5] Weiler M, Sattel S, Giessen T, Jung K, Ehrhardt H, Robertson J. Phys Rev B 1996;53:1594.

[6] Robertson J. Phys Rev B 1996;53:16302.

[7] Robertson J. Phil Mag B 1992;66:199.

[8] Liedtke S, Jahn K, Finger F, Fuhs W. J Non-Cryst Solids 1987;97:1083.

[9] Schutte S, Will S, Fuhs W. Diamond Related Mater 1993;2:1360.

[10] Rusli G, Amaratunga J, Robertson J. App Phys 1996;80, 2998.

[11] Giorgis F, Giuliani F, Pirri CF, Tagliaferro A, Tresso E. App Phys Lett 1998;72:2520.

[12] Ilie A, Conway N, Kleinsorge B, Robertson J, Milne WI. J App Phys 1998;84:5575.

[13] Kleinsorge B, Ilie A, Chhowalla M, Fukarek W, Milne WI, Robertson J. Diamond Related Mater 1998;7:472.

[14] Conway NMJ, Robertson J, Milne WI. Diamond Related Mater 1998;7:477.

[15] Street RA. Hydrogenated amorphous silicon. Cambridge University Press, 1991.

[16] Ristein J, Schafer J, Ley L. Diamond Related Mater 1996;4:508.

Pergamon

Carbon 37 (1999) 835–838

CARBON

Structure, dynamics and optical properties of fullerenes C_{60}, C_{70}

Yu.I. Prilutski[a,*], E.V. Buzaneva[a], L.A. Bulavin[a], P. Scharff[b]

[a]*Dept. of Physics, Kiev Shevchenko University, Vladimirskaya Str., 64, 252033 Kiev, Ukraine*
[b]*Institut fur Anorganise and Analytische Chemie, TU Clausthal, Paul-Ernst-Strase 4, D-38670 Clausthal-Zellerfeld, Germany*

Received 16 June 1998; accepted 3 October 1998

Abstract

The results of a study of structure, dynamic and optical properties of the most stable C_{60}, C_{70} fullerenes and their crystalline phases are reported. The comparison of obtained theoretical data with experimental is made. © 1999 Elsevier Science Ltd. All rights reserved.

Keywords: A. Fullerene; D. Optical properties

1. Introduction

Recently performed experimental studies of physical and chemical properties of fullerenes indicate unique perspectives as regards their future practical use [1]. This necessitate theoretical explanation of the results allowing us to forecast the new effects. First, this concerns a study of structure and mechanical properties of fullerenes and fullerites as a function of temperature and external pressure. Their vibrational spectra need to be studied in detail because the mechanisms of superconductivity and spin-lattice relaxation (SLR) of triplet excited states allowing for low-frequency phonons and intramolecular vibrations [2–4] have remained still actual.

2. Study of vibrational spectra of fullerenes C_{60}, C_{70} and dynamic Jahn-Teller effect in C_{60} fullerene

The calculation of the intramolecular spectra of fullerenes C_{60} and C_{70} was carried out within the framework of a vibrational model using only four independent force constants for the determination of the potential energy of these molecules

$$V = \frac{1}{2}\left(k_1 \sum_j d_{1j}^2 + k_2 \sum_j d_{2j}^2 + k_3 \sum_j \alpha_{1j}^2 + k_4 \sum_j \alpha_{2j}^2 \right)$$

(1)

where k are the force constants which describe the relative changing of the distance (d) between the carbon atoms in pentagons (k_1), hexagons (k_2), and also the changing of the value of angle (α) in pentagons (k_3) and hexagons (k_4) (see Table 1).

The calculated vibrational frequencies of isolated C_{60} and C_{70} molecules are in a good agreement with available experimental results [5,6] in the low temperature range ($T < 25$ K): $\omega_{\text{intra}} = (296–1581)$ cm^{-1} for fullerene C_{60} and $\omega_{\text{intra}} = (241–1570)$ cm^{-1} for fullerene C_{70}. The theoretical values differ from the experimental ones by up to 11%.

For the quantitative analysis of the dynamic Jahn-Teller

Table 1
Calculated values of force constants k_1, k_2 (in kcal/Å^2) and k_3, k_4 (in kcal/(rad)^2)

Force constants	Fullerene C_{60}	Fullerene C_{70}
k_1	1.44×10^{-21}	0.51×10^{-21}
k_2	1.24×10^{-21}	0.60×10^{-21}
k_3	1.09×10^{-6}	0.50×10^{-5}
k_4	0.46×10^{-5}	0.50×10^{-5}

*Corresponding author.

0008-6223/99/$ – see front matter © 1999 Elsevier Science Ltd. All rights reserved.
PII: S0008-6223(98)00280-2

effect in C_{60} fullerene the following approach was used: by changing the geometry of molecule we have minimized its potential energy (see Eq. (1)) as a function of the value of distortion vector (the resulting vector of atom displacements in a molecule) for the corresponding vibrational modes (the obtained results for the excited H_g (437 cm^{-1}) mode are represented in Table 2). As one can see for the split vibrations we have $\omega_{JT} < \omega_{intra}$. The energy of molecule decreases by 23.2 meV. This result is in excellent agreement with the analysis of Jahn–Teller distortions in the C_{60}^- molecular ion [7], where it was found that the energy decreases by 24 meV. The C–C bond length change within the range $(10^{-4}–10^{-3})$ Å and does not exceed the predicted estimate [8] 10^{-2} Å.

3. Study of the structure and vibrational spectra of fullerenes C_{60}, C_{70}

At low temperatures ($T < 25$ K) the equilibrium structure of C_{60} and C_{70} is determined by the minimum of the free energy

$$F = U(\varphi, V) + P\frac{V}{4}, \tag{2}$$

where U is the potential energy of the interaction between the molecules, φ is the setting angle of molecules in the unit cell, $V/4$ is the volume of the unit cell of the crystal per molecule, P is the hydrostatic external pressure.

Table 2
Calculated values of vibrational frequencies (ω_{JT}, cm^{-1}) for the corresponding Jahn–Teller distortion of H_g mode (437 cm^{-1}) in C_{60} molecule

Symmetry	ω_{intra}	ω_{JT}
H_g	296	168
		172
		191
		226
		229
G_u	322	238
		258
		261
		264
F_{2u}	378	272
		275
		287
H_u	401	304
		306
		308
		313
		314
H_g	437	320
		321
		323
		324
		327

The function $U(\varphi, V)$ is described by the atom–atom Lennard–Jones (12-6) potential:

$$U(\varphi, V) = \sum_{\mu,\ \mu'} \left\{ 4\varepsilon \sum_{i,j} \left[\left(\frac{\sigma}{|r_{\mu i} - r_{\mu' j}|} \right)^{12} \right. \right.$$
$$\left. \left. - \left(\frac{\sigma}{|r_{\mu i} - r_{\mu' j}|} \right)^6 \right] \right\} \tag{3}$$

Minimizing the free energy (see Eq. (2)) over the parameters j and V at constant pressure P was carried out in the approximation to rigid molecules under the condition of symmetry conservation of the crystal.

The calculations show that the lattice energy minimum at $P = 0$ is equal to $U_{min} = -1.72$ eV at $\varphi = 23.8°$ (for cubic crystal C_{60}) and $U_{min} = -1.99$ eV at $\varphi = 0$ (for monoclinic crystal C_{70}). The results obtained are in a good agreement with X-ray diffraction data: $\varepsilon = 22°$ at $T = 5$ K (for C_{60} [9]) and $\varphi = 0$ at $T = 10$ K (for C_{70} [10]). The calculated lattice energy for fullerene crystal C_{70} is different from the predicted estimate [11] less than by 5%.

These results were obtained using the following intermolecular potential parameter values (see Eq. (3)) $\varepsilon = 2.964$ meV and $\sigma = 3.47$ Å.

The theoretical-group analysis of the dynamic problem of C_{60} and C_{70} in the center of the Brillouin zone allows one to obtain the analytical expressions for the calculation of nine (for cubic crystal C_{60} [12]) and eleven (for monoclinic crystal C_{70} [13]) fundamental intermolecular vibrational frequencies. These values lie in the following range: $\omega_{inter} = (17–51)$ cm^{-1} for C_{60} and $\omega_{inter} = (13–63)$ cm^{-1} for C_{70}. The theoretical values differe from the experimental data [14,15] by up to 8% at $T \leq 23$ K and $P = 0$.

Over the pressure range used (0 to 1.5 GPa) the librational modes shift linearly towards higher frequencies at a rate of 6.5 cm^{-1}/GPa for C_{60} and 10.2 cm^{-1}/GPa for C_{70}. This result is in a satisfactory agreement with the experimentally observed value for the librational modes [16]: 5.2 cm^{-1}/GPa at $P = 0$ to 1.52 GPa.

The calculated average velocities (in km/s) for longitudinal (v_l) and transverse (v_t) acoustic waves in C_{60} and C_{70} are equal to $v_l = 3.9$, $v_t = 2.3$ and $v_l = 3.0$, $v_t = 1.8$, respectively. They are in a good agreement with the experimental estimate of the average sound velocities in fullerene C_{60} crystal [17]: $v_l = 3.6$, $v_t = 2.1$ at $P = 0$ and $T = 1.4$ to 20 K.

For C_{60} we have found the following values of elastic constants (in GPa): $c_{11} = 22.4$, $c_{12} = 9.1$ and $c_{44} = 11.6$ at $P = 0$. The calculated values of elastic constants for C_{70} $c_{11} = 27.41$ and $c_{12} = 10.94$ are in a satisfactory agreement with the experimental estimate [18]: $c_{11} = 30.87$ and $c_{12} = 13.12$ at $P = 0$.

These results were obtained using the following intermolecular potential parameter values (see Eq. (3)) $\varepsilon = 0.059$ meV and $\sigma = 3.47$ Å.

4. Study of spin-lattice relaxation in the excited triplet states of fullerenes C_{60}, C_{70}

We have obtained the following expressions for the SLR transition probabilities between the triplet spin sublevels ($\bar{1}$, 0, 1) of the photoexcited molecules C_{60} and C_{70} caused by the direct one-phonon processes: in the framework of TRM mechanism (the mixing of the translational and rotational motions of a molecule [3])

$$W_{10}^{(TRM)} = aB^3 \left[1 - \exp\left(-\frac{g\beta B}{kT} \right) \right]^{-1} (\cos^2 \varphi + \cos^2 2\varphi), \tag{4}$$

$$W_{1\bar{1}}^{(TRM)} = 16aB^3 \left[1 - \exp\left(-\frac{g\beta B}{kT} \right) \right]^{-1} \sin^2 \varphi (1 + \cos^2 \varphi), \tag{5}$$

where φ is the angle between the magnetic field vector B and the X molecular axis. The spin-phonon interaction constant is equal to $a = 2.98 \times 10^2$ s^{-1}T^{-3} (in the case of C_{60}) and $a = 1.73 \times 10^2$ s^{-1}T^{-3} (in the case of C_{70}); in the framework of TVM mechanism (the mixing of the translational motions of a molecule and vibrational motions of nuclei [4])

$$W_{10}^{(TVM)} = \tilde{a}B^5 \left[1 - \exp\left(-\frac{g\beta B}{kT} \right) \right]^{-1} \sin^2 2\varphi, \tag{6}$$

$$W_{1\bar{1}}^{(TVM)} = 64\tilde{a}B^5 \left[1 - \exp\left(-\frac{2g\beta B}{kT} \right) \right]^{-1} \sin^4 \varphi, \tag{7}$$

where the spin-phonon interaction constant is equal to $\tilde{a} = 3.30$ s^{-1}T^{-5} (in the case of C_{60}) and $\tilde{a} = 3.78$ s^{-1}T^{-5} (in the case of C_{70}).

As one can see from Table 3 the TRM mechanism dominates over the TVM mechanism. But in the high magnetic field range $B \geq 4.8$ T the TVM mechanism begins to dominate, since $(W^{(TVM)}/W^{(TRM)}) \sim B^2$. Finally, the contribution of the TVM mechanism to the SLR can

essentially increase due to the existence of dynamic Jahn-Teller effect in C_{60} fullerene.

The numerical values of SLR probability in the triplet excited states of C_{70} in the range of magnetic fields $0.3 \leq B \leq 5$ T and $T = 4.2$ K at $\varphi/\pi = p/2$ are equal to $W_{1\bar{1}}^{(TRM)} = (4.28 \times 10^2 - 3.61 \times 10^5)s^{-1}$ and $W_{1\bar{1}}^{(TVM)} = (3.36 - 7.88 \times 10^5)s^{-1}$.

We have obtained the following expressions for the SLR transitions probability in the triplet photoexcited molecule C_{60} caused by the Raman two-phonon processes

$$W_{10}^{(TRM)} = \frac{16\pi^6}{21} bT^7 (22 - 3\cos^2 \varphi - 17\cos^2 2\varphi), \tag{8}$$

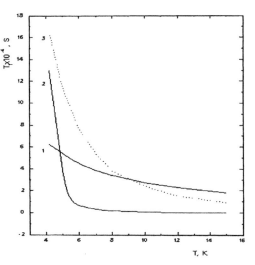

Fig. 1. Dependence of SLR time on the temperature: curve 1 and 2 describe the direct one-phonon and Raman two-phonon relaxation processes, respectively; curve 3 shows the experimental data for the paramagnetic centers of polycrystalline C_{60} [19].

Table 3
Calculated values of SLR probability $W_{1\bar{1}}$ (in s^{-1}) in the triplet excited states of fullerene C_{60} for different temperatures and magnetic fields ($B \| Z$), obtained in the framework of the TRM and TVM mechanisms. The values of SLR probability in consideration of Jahn-Teller effect influence are represented in brackets.

SLR probability	B, T	T, K			
		4.2	10	15	20
$W_{1\bar{1}}^{(TRM)}$		7.36×10^2	1.66×10^3	2.46×10^3	3.26×10^3
$W_{1\bar{1}}^{(TVM)}$	0.3	2.94	6.62	9.80	130×10^1
		(1.41×10^1)	(3.17×10^1)	(4.69×10^1)	(6.22×10^1)
$W_{1\bar{1}}^{(TRM)}$		1.01×10^4	2.02×10^4	2.90×10^4	3.79×10^4
$W_{1\bar{1}}^{(TVM)}$	1	4.47×10^2	8.95×10^2	1.29×10^3	1.68×10^3
		(1.88×10^3)	(3.76×10^3)	(5.42×10^3)	(7.06×10^3)
$W_{1\bar{1}}^{(TRM)}$		6.21×10^5	8.06×10^5	1.01×10^6	1.22×10^6
$W_{1\bar{1}}^{(TRM)}$	5	6.88×10^5	8.93×10^5	1.11×10^6	1.35×10^6
		(1.65×10^6)	(2.14×10^6)	(2.66×10^6)	(3.24×10^6)

838

Y.I. Prilutski et al. / Carbon 37 (1999) 835–838

$$W_{1f}^{(TRM)} = \frac{16\pi^6}{21} bT^7 (21 - 3\cos^2\varphi - 7\cos^2 2\varphi), \qquad (9)$$

where the spin-phonon interaction constant is equal to $b = 1.28 \times 10^{-6} \text{ s}^{-1} \text{ K}^{-7}$.

As one can see from Fig. 1 at the helium temperature the Raman two-phonon (curve 2) processes are dominant over the direct one-phonon (curve 1) ones. When the temperature is increased ($T > 5$ K) the one-phonon processes begin to dominate. The calculated curves 1 and 2 are in a satisfactory agreement with the experimental result [19] (curve 3) in the temperature range of $T = (4.2–15)$ K: the profiles of curves 2 and 1 copy the curve 3 in the range of $T = (4.2–5)$ K and $5 < T \leq 15$ K, respectively.

References

[1] Eickenbusch H, Hartwich P. Fullerene (Technologie-Analyse). VDI Technologie-Zentrum, Physikalische Technologien, 1993.
[2] Chen G, Goddard WA. Proc Nat Acad Sci (USA) 1993;90:1350.
[3] Andreev VA, Prilutski YuI. Mol Cryst Liq Cryst 1996;289:255.
[4] Andreev VA, Prilutski YuI. Mol Cryst Liq Cryst 1997;307:205.
[5] Christides C, Nikolayev AV, Dennis TJS, et al. J Chem Phys 1993;97:3631.
[6] Schettino V, Salvi PR, Bini R, Cardini G. J Chem Phys 1994;101:11079.
[7] de Coulon V, Martins JL, Reuse F. Phys Rev B 1992;45:13671.
[8] Asai Y, Kawaguchi Y. Phys Rev B 1992;46:1265.
[9] David WIF, Ibberson RM, Matthewman JC, et al. Nature 1991;353:147.
[10] Janaki J, Rao GVN, Sastry VS, et al. Sol St Commun 1995;94:37.
[11] Verheijen MA, Meekes H, Meijer G, et al. Chem Phys 1992;166:287.
[12] Prilutski YuI, Shapovalov GG. Phys Status Sol (b) 1997;201:361.
[13] Prilutski YuI, Andreev VA, Shapovalov GG. Czech J Phys 1996;46:2711.
[14] van Loosdrecht PHM, Verheijen MA, Meekes H, van Bentum PJM, Meijer G. Phys Rev B 1993;47:7610.
[15] Horoyski PJ, Thewalt MLW. Phys Rev B 1993;48:11446.
[16] Horoyski PJ, Wolk JA, Thewalt MLW. Sol St Comun 1995;93:575.
[17] Beyermann WP, Hundley MF, Thompson JD, Diederich FN, Gruner G. Phys Rev Lett 1992;68:2046.
[18] Callebaut AK, Michel KH. Phys Rev B 1995;52:15279.
[19] Zaritskii IM, Ischenko SS, Konchits AA, et al. Fiz Tverd Tela 1996;38:419.

Pergamon

Carbon 37 (1999) 839–842

CARBON

Doping mechanism in tetrahedral amorphous carbon

C.W. Chen, J. Robertson*

Engineering Department, Cambridge University, Cambridge CB2 1PZ, UK

Received 16 June 1998; accepted 3 October 1998

Abstract

Doping in hydrogenated amorphous silicon occurs by a process of an ionised donor atom partially compensated by a charged dangling bond. The total energies of various dopant and dopant/bonding combinations are calculated for tetrahedral amorphous carbon. It is found that charged dangling bonds are less favoured because of the stronger Coulombic repulsion in ta-C. Instead the dopants can be compensated by weak bond states in the lower gap associated with odd-membered π-rings or odd-numbered π-chains. The effect is that the doping efficiency is low but there are not charged midgap recombination centres, to reduce photoconductivity or photoluminescence with doping, as occurs in a-Si:H. © 1999 Elsevier Science Ltd. All rights reserved.

Keywords: A. Amorphous carbon; B. Doping; D. Photoconductivity

1. Introduction

The doping mechanism of amorphous semiconductors has always been an interesting issue. In the early days of amorphous semiconductors, it was believed that doping would not occur because the absence of long range order would allow potential dopant atoms to exert their chemically preferred valence, which was a non-doping or 'alloying' state [1]. For hydrogenated amorphous silicon (a-Si:H) this meant that a phosphorus atom would prefer a trivalent P_3^0 site rather than a four-fold substitutional P_4^0 site. (Here, the subscript denotes the coordination and the superscript the site's formal charge). Then Spear and LeComber discovered doping of a-Si:H [2]. The doping was inefficient because most phosphorus atoms still occupied the alloying P_3^0 site and a small fraction occupied the substitutional P_4^0 site. It was later realised that the doping mechanism was more complex than this [3,4]. Instead of phosphorus occupying the simple P_4^0 site, it was ionised with its electron going to a created dangling bond site to form a negatively charged defect, to give P_4^+-D^- pair. This site-pair is compensated, but it still provides doping because it pushes the Fermi level above the defect level into the upper gap. The presence of extra charged

defects in doped a-Si:H is detrimental to its electronic properties such as photoluminescence or carrier lifetimes, which are found to degrade in doped material [5]. The various impurity configurations are shown in Fig. 1 for the carbon equivalents.

Doping of diamond-like carbon has also been contentious. The early hydrogenated amorphous carbon (a-C:H) prepared by plasma enhanced chemical vapour deposition (PECVD) is not easily doped. Meyerson and Smith [6] showed the change of thermopower expected for bipolar

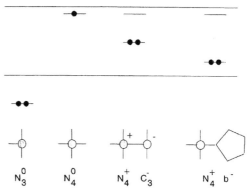

Fig. 1. Various configurations and associated gap states of N in amorphous carbon.

*Corresponding author. Tel: 44 1223 332689; fax 44 1223 332662.
E-mail address: jr@eng.cam.ac.uk (J. Robertson)

0008-6223/99/$ – see front matter © 1999 Elsevier Science Ltd. All rights reserved.
PII: S0008-6223(98)00281-4

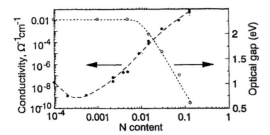

Fig. 2. Experimental variation of conductivty and optical gap with nitrogen addition to ta-C [9–11].

doping, and Jones and Stewart [7] showed doping by nitrogen, but others [8] argued that the observed conductivity increased were mainly due to a gradual dopant-induced graphisation. A clearer doping effect was apparent in so-called tetrahedral amorphous carbon (ta-C) which is a hydrogen-free a-C with a high fraction of sp3 sites. The ta-C can be formed by various processes including the filtered cathodic vacuum arc (FCVA) or pulsed laser deposition. Veerasamy et al [9], Davis et al [10] and McKenzie et al [11] have shown nitrogen doping of ta-C. The undoped ta-C appears slightly p-type, so the conductivity first decreases then increases as the Fermi level moves across the gap (Fig. 2). It is clear than the doping efficiency is even weaker than in a-Si:H, partly because the ta-C possesses a considerable intrinsic defect density of order 10^{19} cm^{-3}. High nitrogen contents induce graphitisation and the optical gap is seen to decline (Fig. 2). Kleinsorge et al [12] have recently found weak boron doping of ta-C.

It is appreciated that a far wider range of configurations of nitrogen is possible in a-C than in a-Si:H because carbon and nitrogen can each occupy sp3 or sp2 sites [13]. It is also interesting to define the cause of the nitrogen induced-graphitisation at high N contents, as this is relevant to the search for the high-modulus C_3N_4 phase, in which C must retain its sp3 bonding [14].

The various nitrogen sites can be classified in terms of whether N is σ or π bonded. For σ bonding, nitrogen can occupy the alloying N_3^0 site (as in C_3N_4), the substitutional N_4^0 site, or the N_4^+-C_3^- pair. If the N has π bonds, the N can be in chain or aromatic groupings. The chain sites are probably the most important sites for N in a-C:H. These are the $=$C$=$N- and the -C$\equiv$N nitrile site [15]. Both these sites are non-doping. For aromatic groupings, the N can be at pyridine-like or pyrrole-like non-doping sites, or at a substitutional site in a benzene ring which is doping. Weich et al [16] have shown that many of these sites have similar total energy, which accounts for why such a range of nitrogen configurations will be observed in a-C systems.

2. Calculations

This paper concerns the nature of the doping configuration itself. We have calculated the total energy of nitrogen at various configurations in carbon. Rather than use N atoms embedded in periodic cells of a-C as others [16–18], we carried out the calculations on simple clusters of N surrounded by C sites, with the bonds at the cluster surface terminated by hydrogen. (Clusters are not a large disadvantage for carbon systems because of the strong localization of the bonding). The total energies were calculated using the total energy code 'Fast Structure' of Molecular Simulations Inc. It has a numerical basis set, it can have a double-zeta equivalent flexibility on each site needed for proper representation of carbon bonding in ta-C systems [19], and it uses the Harris functional and local density approximation to find the total energy terms. The total energies are displayed in Fig. 3, referenced to a zero defined for the N_3^0 site surrounded by sp3 carbons. The referencing was carried out between clusters by subtracting group bonding energies found from the calculations.

Figure 3 shows the total energy of some of these nitrogen configurations. It is seen that the N_3^0 site has the lowest energy. The non-doping C$=$N group is very stable with an energy 0.68 eV, while the chain terminating C$\equiv$N nitrile group has an energy of about 1.1 eV. The other non-doping site shown, the 5-fold sp2 ring with a N site (pyrrole) also has quite a low energy of 0.43 eV. Thus there are a number of alternative, quite stable non-doping configurations for N in a-carbons, as noted previously [16].

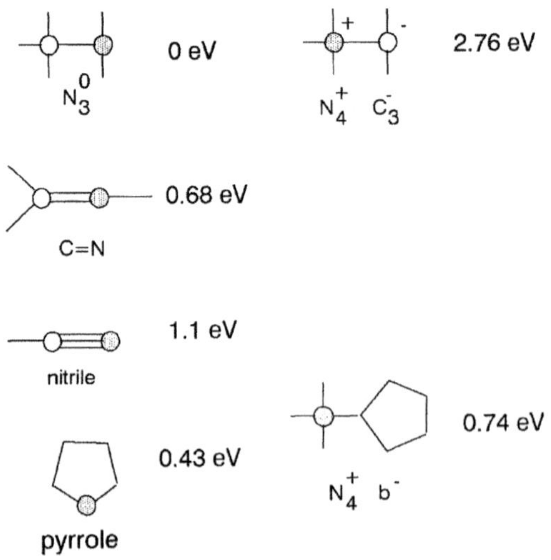

Fig. 3. Calculated total energies of various N configurations in carbon, referred to the N_3^0 site.

The N_4^+-C_3^- site pair is seen to have a rather high energy of 2.76 eV. The simple substitutional site N_4^0 has an even higher energy, and is unstable to bond breaking in small clusters, but not in networks where it is constrained. The high energy of N_4^+-C_3^- suggests that this will not be a prevalent site in ta-C.

The presence of π bonding allows for other compensating configurations not considered previously. A 5-fold π-bonded ring has five eigenstates, a doubly occupied state at 2β, an empty 2-fold degenerate state at -1.62β and a state at $+0.6\beta$ which is partly filled with 3 out of a possible 4 electrons [20]. This partly filled state destabilises isolated 5-fold sp2 rings under normal circumstances. However, in the presence of donors, it can accept an electron to act as a compensating site. A similar configuration with a partly filled, weak π bonded state occurs in odd-numbered olefinic chains.

The N_4^+/5-fold ring pair denoted N_4^+-b^-, (b for bond), is seen to have a relatively low total energy of 0.74 eV, much lower than the N_4^+-C_3^- pair. This N_4^+-b^- still gives doping, because the Fermi level moves up in the gap. This suggests that N_4^+-b^- could be the preferred configuration of dopants in ta-C (or an equivalent chain configuration).

Why is the N_3^0 site not so stable? We believe that this is because of the greater orbital localisation in carbon systems which destabilises the carbanion C_3^-. For example, the correlation energy, U, of the dangling bond defect in a-Si:H is about 0.2 to 0.3 eV [5]. In a-C:H, it is estimated from the ratio of the spin density to the density of states at the Fermi level from photoemission that U is about 2 to 3 eV [21]. This is a sizeable destabilisation of the C_3^- site. The second electron of C_3^- is strongly localised on the central C site, whereas in Si it is more delocalized onto its neighbours. In contrast, for the N_4^+-b^- configuration, the negative charge is in a π state delocalized over more sites. This lowers its U.

The N_4^+-b^- pair has two gap states, the upper donor state due to N and the state in the lower gap due to the ring (Fig. 1). This state is in the lower part of the gap because it is partly bonding.

3. Implications

It is possible that this lower position, almost lying in the valence band will have a significant effect on the behavior of doped ta-C and ta-C:H. As noted above, the charged midgap levels introduced by doping in a-Si:H have a strong effect on its electronic properties. The luminescence efficiency and the $\mu\tau$ product of majority carriers both decrease quickly with doping. This is a significant effect and requires a-Si:H based solar cells to use a p-i-n structure. Midgap levels are efficient recombination centres because they allow the energy to be lost to phonons in two stages [5]. On the other hand, a level near the valence band

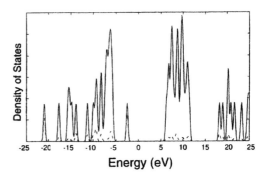

Fig. 4. Density of states of N_3^0 site.

is not such a strong recombination centre as a midgap level, so the N_4^+-b^- pair may not be a strong recombination centre. This is supported by the preliminary observation that the luminescence efficiency in ta-C:H does not decline with nitrogen addition [22].

4. Density of states

Figures 4–7 show the calculated density of states (DOS) for nitrogen and the surrounding carbon atoms for some of these configurations. The N_3^0 site is nondoping because its 3 valence p electrons fill the bond states and its 2 s electrons fill a lone pair, at -3 eV in Fig. 4. The N_4^0 site would have 4 electrons in four sp3 bonds and the fifth electron in an antibonding state which becomes the shallow state just below the conduction band edge. In the N_4^+-C_3^- pair, the electron has dropped into the defect state at midgap. This lies at 0 eV in Fig. 5. Fig. 6 shows the DOS of the N_4^+-b^- pair. The empty donor state is at 3 eV

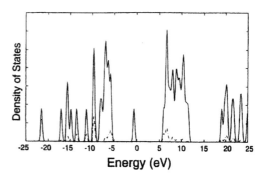

Fig. 5. Density of States of the N_4^+-C_3^- pair.

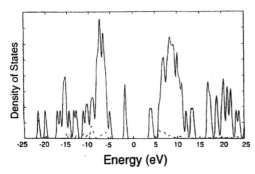

Fig. 6. Density of states of the N_4^+-b^- pair.

and its electron has dropped into the π weakly bonding state at -2 eV, which is the valence band. Figure 7 shows the DOS of the -C≡N group with its wide empty gap, characteristic of a non-doping group.

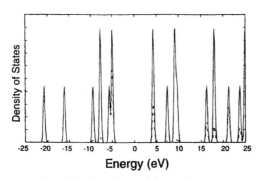

Fig. 7. Density of states of the -C≡N group.

References

[1] Mott NF. Adv Phys 1967;16:49.
[2] Spear WE, LeComber PG. Adv Phys 1977;26:811.
[3] Street RA. Phys Rev Lett 1982;49:1187.
[4] Robertson J. Phys Rev B 1985;31:3817.
[5] Street RA. Hydrogenated amorphous silicon. Cambridge University Press, Cambridge, 1991.
[6] Meyerson B, Smith FW. Solid State Commun 1982;41:68.
[7] Jones DI, Stewart AD. Philos Mag B 1982;46:423.
[8] Helmbold A, Hammer P, Thiele JU, Rohwer K, Meissner D. Phil Mag B 1995;72:335.
[9] Veerasamy VS, Yuan J, Amaratunga G, et al. Phys Rev B 1993;48:17954.
[10] Davis CA, McKenzie DR, Yin Y, Kravtchinskaia E, Amaratunga GAJ, Veerasamy VS. Phil Mag B 1994;69:1133.
[11] McKenzie DR, Yin Y, Marks NA, et al. J Non-Cryst Solids 1993;164:1101.
[12] Kleinsorge B, Ilie A, Chhowalla M, Fukarek W, Milne WI, Robertson J. Diamond Relat Mater 1998;7:472.
[13] Robertson J, Davis CA. Diamond Relat Mater 1994;4:441.
[14] Liu AM, Cohen ML. Phys Rev B 1990;41:10727.
[15] Kaufman H, Metin S, Saperstein DD. Phys Rev B 1989;39:13053.
[16] Weich F, Widany J, Frauenheim T. Phys Rev Lett 1997;78:3326.
[17] Frauenheim T, Jungnickel G, Sitch P, et al. Diamond Relat Mater 1998;7:348.
[18] Stumm P, Drabold DA, Fedders PA. J App Phys 1997;81:1289.
[19] Nelson JS, Stechel EB, Wright AF, Plimton SJ, Schultz PA, Sears MP. Phys Rev B 1995;52:9354.
[20] Robertson J. Adv Phys 1986;35:317.
[21] Ristein J, Schafer J, Ley L. Diamond Relat Mater 1995;4:508.
[22] Silva SRP. Private communication.

Pergamon

Carbon 37 (1999) 843–846

 **CARBON**

Chromium in amorphous hydrogenated carbon based thin films prepared in a PACVD process

P. Gantenbein[a,*], S. Brunold[a], U. Frei[a], J. Geng[b], A. Schüler[b], P. Oelhafen[b]

[a]School of Engineering ITR, Oberseestr. 10, CH-8640 Rapperswil, Switzerland
[b]University of Basel, Klingelbergstr. 82, CH-4056 Basel, Switzerland

Received 16 June 1998; accepted 3 October 1998

Abstract

The deposition of Cr-containing amorphous hydrogenated carbon (a-C:H) thin films on Si and Cu substrates in a radio frequency (rf) and middle frequency (mf) plasma activated chemical vapour deposition (PACVD) process is described. Negative DC bias voltage at the substrate and the composition of the process gas are two significant parameters in the deposition process. By applying a higher bias voltage the growth rate is increased. The dependence on the ratio of CH_4 to Ar shows a saturation effect in the region of high CH_4 concentration. This shows the significant contribution of an ion current to the film growth and a rf/mf power limited ion current in the selected parameter range. A decrease of the Cr content in the a-C:H films by increasing the CH_4 to Ar flow ratio is measured by XPS while a correlation of the density of the electronic states at the Fermi level with the metal content in the a-C:H matrix is observed by UPS. No significant change in the ratio of sp^2 to disordered bonded carbon atoms in the films deposited with the two methods is observed by Raman spectroscopy. Ageing tests of the samples at 250°C in air showed the increase of the solar absorption as well as the reduction of the thermal emission after 80 h treatment. These results show a total suppression of the copper diffusion through the diffusion barrier in the multilayer system. © 1999 Elsevier Science Ltd. All rights reserved.

Keywords: A. Amorphous carbon; Thin films; B. Sputtering; D. Electronic structure

1. Introduction

Recent development in the market of solar energy conversion has become more and more interesting for photovoltaic as well as for solar thermal application. In previous work an application of chromium containing a-C:H films for solar thermal application in a flat plat absorber has been shown [1]. These coatings are deposited in a process that combines plasma activated chemical vapour deposition (PACVD) of methane and rf 13.56 MHz magnetron sputtering of a chromium target with Argon. The resulting films are cermets of hydrogenated amorphous carbon and chromium carbide. In this process a proper adjustment of the plasma impedance to the power supply and the connecting cables by a matching network is essential for power flow into the plasma. To overcome the subtle problem of rf power consumption in the matching network a lower excitation frequency in the process is

desirable. The purpose of this work is to show the frequency dependence of the structure of the a-C:H/Cr films. The morphology of these films is a decisive parameter for the application in thermal solar energy conversion.

2. Experimental

The schematic of the deposition configuration for preparing a-C:H/Cr coatings is shown in Fig. 1 of [1] where the experimental procedure is also described. In this work we give a short supplementary description to the previous paper. Flow controllers were used to feed argon of 3.5 to 16 sccm and methane of 1.5 to 7.0 sccm independently into the chamber. The working pressure of 2–4 Pa was adjusted by a pressure controller valve. The substrates were resistively heated to the temperature of 200°C. The 40 kHz power of 200 W was coupled by a step-down/step-up transformer to the target carrying magnetron, yielding a typical target DC self-bias of about −430 V. The peak to

---

*Corresponding author.

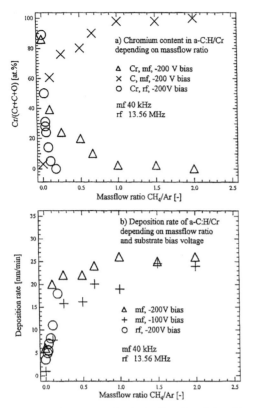

Fig. 1. Chromium concentration in the films (a) and deposition rate on silicon substrates (b) depending on the process gas mixture in the rf/mf magnetron sputter CVD process.

peak mf voltage measured by an oscilloscope at this power level was about 2000 V. The chromium concentration of the samples was determined by XPS (Mg K$_\alpha$, $h\nu=1253.6$ eV). The valence band structure of the a-C:H/Cr films was measured by UPS (HeI, $h\nu=21.22$ eV). The deposition rate depending on the process gas mixture was determined by an α-step. Before deposition of a-C:H films on silicon wafer substrates the plasma was run for a minimum of 10 min with a high concentration of methane in the process gas to be sure a total carbon based cover of the metal target was reached. These samples were characterised by Raman spectroscopy. The multilayer absorber coatings deposited on Cu substrates have been subjected to accelerated ageing at an elevated temperature of 250°C in air. Before and after the treatment, the total reflectance of the samples in the wavelength range from 0.35 μm to 18 μm was measured with a Bruker FTIR spectrometer. The solar absorbance (Air Mass 1.5), α_s, and thermal emittance at 100°C, $\varepsilon_{100°C}$, were calculated from the total reflectance at room temperature.

3. Results and discussion

Figure 1 shows the Chromium concentration in the films and the deposition rate of a-C:H/Cr on silicon substrates depending on the process gas mixture. The chromium concentration in the a-C:H films can be varied by varying the CH$_4$ to Ar gas ratio in the rf as well as in the mf mode. To prepare films in this mode without Cr in the dielectric a-C:H matrix, a much higher ratio of CH$_4$ to Ar is needed. The deposition rate is strongly increased by increasing the ratio of methane in the process gas mixture. While the film growth rate reaches a saturation value of about 25 nm/min at a high amount of methane in the 40 kHz excited plasma, the deposition rate is remarkably influenced by the substrate bias voltage in the gas ratio range of 0.1 to 1.0 of methane to argon. By applying a bias of -200 V the growth rate is significant higher than at a bias of -100 V. This effect can be explained by the contribution to the deposition of a higher amount of ions and radicals formed out of methane [2]. For comparison the values of the 13.56 MHz plasma excitation are indicated in Fig. 1. While the deposition rate in the rf mode is smaller compared to the 40 kHz mode a higher amount of a-C:H phase is formed by a lower fraction of CH$_4$ in the process gas. The formation of the dielectric a-C:H material is reached in the 40 kHz mode by a higher fraction of methane compared to argon. In the 40 kHz mode the chromium concentration in the a-C:H/Cr films is higher than in the 13.56 MHz mode at the same methane fraction because of a temporary much higher target bias voltage. A higher target bias leads to an increased sputtering yield on the chromium target surface by the increased energy of the impinging argon ions [3]. In the mf mode of 40 kHz the ions follow the electric mf field and the sputtering yield at the temporal average target bias of up to -1400 V is twice as the yield in the rf mode with a target bias of -435 V. In the mf excitation a methane to argon fraction in the process gas of minimum 2/1 is needed for chromium free formation of a-C:H matrix. At these process gas mixtures a net positive film growth rate on the target starts poisoning the metal target.

The UPS He I data of the mf a-C:H/Cr films are shown in Fig. 2. The spectra show an increased density of states (DOS) near the Fermi energy E_F for an increased chromium content in the films. In the spectra two distinct regions are visible. In the energy range near the Fermi energy the 3d states of Cr are dominating the spectra while at low Cr content the σ peak of a-C:H is strongest. In the energy range of 3.0 eV to 5 eV the DOS is dominated by a hybridisation peak of C2p and Cr3d states. The spectra of a-C:H consists of the σ peak and the π shoulder near the Fermi energy indicating sp^2 co-ordinated carbon atoms [4]. The formation of a chromium carbide phase is shown in the XPS C1s core level spectra of Fig. 2. The concentration of this carbide in the films is increased by an increasing amount of Cr. This chromium carbide can be identified by Cr$_7$C$_3$ which is revealed by XRD. This XRD

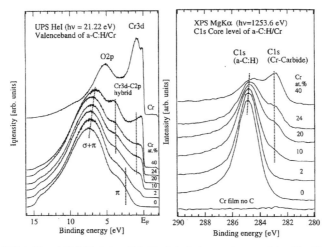

Fig. 2. XPS (Mg K$_\alpha$, $h\nu$=1253.6 eV) and UPS (He I, $h\nu$=21.2 eV) photoelectron spectroscopy data of the C1s core level and the valence band of the a-C:H/Cr films.

data as well as the SEM measurements will be presented elsewhere. Even for high metal concentrations in the films a certain amount of a-C:H phase is present and plays an important role in the application of this composite material as a diffusion barrier for copper. For comparison of the atomic structure of a-C:H prepared at the two different excitation frequencies of 13.56 MHz and 40 kHz, Raman spectra are shown in Fig. 3. The spectra are dominated by the well known G-peak at 1541 cm^{-1} and D-peak/shoulder at 1331 cm^{-1} relative to the main peak of the argon ion laser at 514.5 nm wave length. In the three spectra of the a-C:H films prepared by rf with substrate bias and prepared by mf with and without bias voltage no difference in the line shape is visible. While the G-line of the a-C:H film deposited in the rf mode without substrate bias is shifted to a higher value of 1566 cm^{-1}. Dielectric a-C:H films prepared in the two plasma excitation modes show no significant difference in the bulk sp^2 (G-peak) and disordered (D-peak) co-ordination of the carbon atoms. We believe this can be explained by the same ion growth mechanism of the films as well as insensitive structure formation in a certain energy range of the ions impinging on the growing film surface. For a high sp^3 bond fraction of carbon atoms in a-C:H films the energy of the arriving ions should be in the range of 100 eV to about 300 eV [5] [6].

Based on these results we conclude that there is no difference in the morphology of the films prepared by the two different plasma excitation methods.

In order to apply these films in a multilayer assembly with a minimum of three layers with different metal concentrations, we measured the reflectance $R(\lambda)$ in the wave length range of 0.35 to 18 μm. For use as solar

selective films in solar thermal conversion the a-C:H/Cr films have to withstand a corrosive atmosphere where humid air plays the dominant role as a destructive agent. The ageing behaviour of these films deposited on the widely used copper substrates for solar application was tested in air at an elevated temperature of 250°C. The

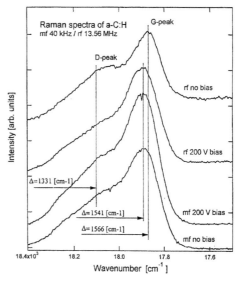

Fig. 3. Raman spectra of a-C:H films prepared by mf and rf excited plasma deposition. For the rf −200 V bias and mf −200 V and mf no bias no significant difference in the spectra shape is visible.

reflectance spectra show a total suppression of copper diffusion to the top of the multilayer for a diffusion barrier of a-C:H/Cr with high Cr content.

4. Conclusions

Cermet films like a-C:H/Cr can be prepared by mf- and rf-magnetron sputter chemical vapour deposition with a process gas mixture of CH_4 and Ar. To achieve the same chromium concentrations in the composite layers a higher fraction of methane in the process gas mixture is needed in the 40 kHz excitation mode. This effect can be explained by a higher average target bias voltage resulting in a higher sputter yield on the chromium target surface in the low frequency mode. The growing material has the same electronic and atomic structure independent of the excitation frequency. Chromium metal is built in the material in a metallic chromium carbide phase. The carbide particles are embedded in a dielectric a-C:H matrix. This is the key morphology for the application as a diffusion barrier on copper substrates. The ageing behaviour of a multilayer film on a copper substrate improves the performance of the solar absorber. Substrate temperature and substrate bias voltage are the decisive parameters in the formation of this composite material in the rf mode. While in the mf mode the high mf voltage amplitude determines the average energy for the ions contributing to the film growth.

Acknowledgements

Financial Support from the Swiss Bundesamt für Energiewirtschaft is gratefully acknowledged. We gratefully acknowledge XRD measurements by Dr. H. P. Lang and J. P. Ramseyer of the University of Basel and SEM measurements by D. Mathys and M. Düggelin of the REM-Laboratory of the University of Basel.

References

[1] Gampp R, Oelhafen P, Gantenbein P, Brunold S, Frei U, Accelerated ageing tests of chromium containing amorphous hydrogenated carbon coatings for solar collectors. In: International Symposium on Optical Materials Technology for Energy Efficiency and Solar Energy Conversion XV, Freiburg, Germany, Sept. 1996.

[2] Keudell A. Growth Mechanisms during the plasma enhanced chemical vapor deposition of hydrocarbon films, investigated by in situ ellipsometry, IPP 9/110, Max Planck-Institut für Plasmaphysik, Garching, 1996.

[3] Behrisch R, editor, Andersen HH, contributor. Sputtering by particle bombardment/sputtering Yield measurements. In: Springer-Topics in Applied Physics, 1981-1991.

[4] Ugolini D, Oelhafen P, Wittmer M. In: Koidl and Oelhafen P, editors, Amorphous hydrogenated carbon films. Les Ulis (France): Les Editions de Physique, 1987:297.

[5] Schäfer J, Ristein J, Ley L. Electronic density of states and deep defects of hydrogenated amorphous carbon (a-C:H). In: Diamond Films, Portugal 1993.

[6] Hofsäss H, Feldermann H, Merk R, Sebastian M, Ronning C. Cylindrical spike model for the formation of diamondlike thin films by ion deposition. Appl Phys A 1998;66:153.

Pergamon

Carbon 37 (1999) 847–850

CARBON

Frenkel-excitations of C_N ($N=12,60$) clusters

Stan F. Kharlapenko, Slava V. Rotkin[*]

Ioffe Physico-Technical Institute, 26, Politehnicheskaya st., 194021 St. Petersburg, Russia

Received 16 June 1998; accepted 3 October 1998

Abstract

The simple Frenkel-exciton model Hamiltonian is applied for calculation of spectrum of electron-hole excitations in carbon nanoclusters. The group-theoretical approach allows to find analytically mode frequencies as well as wavevectors of excitations. © 1999 Elsevier Science Ltd All rights reserved.

Keywords: A. Carbon clusters; D. Electronic structure

1. Introduction

The finite carbon nanoscale cluster electronic system is still a challenge to theoreticians. It has the intermediate scale: neither atomic nor 3D-bulk one, which troubles the typical many-body method application. There is some theoretical and experimental evidence that Coulomb interaction should be taken into account for these systems. Some interesting analytical results were obtained [1] for the fullerenes within SO(3)-spherical symmetry approximation – though the actual icosahedral symmetry group is much lower than the infinite full group of rotations. We note that the symmetry of the tight-binding (TB) one-electron Hamiltonian is high enough to obtain, for example, the spectrum within purely analytical methods [2–4].

The high symmetry leads us to consider some simplified Coulomb Hamiltonian for C_{60}-like lattice of atoms possessing one electron-hole chargeless excitation per site. The Wigner–Ekkart theorem allows to expand all operators into irreducible operator series for any lattice (here the icosahedral 60-membered lattice is considered). We write the Hamiltonian in the standard secondary quantization (electron-hole) formalism. The next step is obvious – the multipole expansion in the first non-vanishing order for the electron-hole excitation is the dipole approximation. When considering only next-neighbor interaction, it is the simplest Frenkel-exciton Hamiltonian well known from organic insulator solid state theory. To be noted, the problem is easily solved numerically. Then we make use of TB

dipole–dipole approximation. One can expect that a number of degrees of freedom is much larger than for translationaly invariant systems, which makes an analytical solution difficult if not possible. Even so, we have shown that the group-theoretical approach gives the exact result for some modes [5]. For example, we calculated analytically the triply-degenerate optically active excitons and non-degenerate excitations of C_{60} and modes of C_{12} (three are dipole active ones).

2. Symmetry classification

A number of carbon nanoclusters of high symmetry have been synthesized over the last years. Along with clusters of perfect geometrical shape, a lot of less symmetrical closed (and even opened) macromolecules of carbon occur in the carbon soot. The main feature of these constructions is a three-coordinated network of carbon–carbon bonds that allows us to treat them as graphite-like surfaces [6]. The most familiar C_{60} cluster has an icosahedral (Y_h) symmetry, being a perfect truncated icosahedron. This network (we will also call it the cluster lattice bearing in mind the correspondence to infinite lattice of 2D graphite sheet) can be obtained as some projection of a triangular group (2,3,5). The notation reflects the symmetry of a basic triangular patch of clusters which has to be rotated by all Y_h group elements to cover all of the surface (see Ref. [7] for more details). The last number 5 in the notation shows that C_{60} has a 5-fold axes (or 5-membered rings). One can consider polyhedrons with 3- and 4-fold axes which give more examples of carbon clusters with a graphite-like curved surface. The first is a

[*]Corresponding author. Tel.: +7-812-247-9367; fax: +7-812-247-1017.
E-mail address: rotkin@theory.ioffe.rssi.ru (S.V. Rotkin)

0008-6223/99/$ – see front matter © 1999 Elsevier Science Ltd All rights reserved.
PII: S0008-6223(98)00283-8

truncated tetrahedron which was considered theoretically as C_{12}. The last one represents the truncated cube and will not be considered below. It has only even-membered rings in the lattice which makes its symmetry slightly different.

We have to note that considered C_{12} and C_{60} clusters are the smallest representatives of the truncated tetrahedron and icosahedron family which is formed by substitution of some graphite lattice fragment instead of a single carbon atom in the same lattice (cf. Ref. [8]). The unified approach allows us to calculate the TBA spectrum of these clusters. The matrix TBA Hamiltonian can be expanded in irreducible representation (IRR) space. The high symmetry results in essential simplification of secular equations, which are solved exactly for clusters with a simple lattice basis [2–4,7,8]. Let us consider the Hamiltonian symmetry of C_{12} cluster in more details.

The TBA Hamiltonian has only two parameters to calculate the one-electron spectrum. There is a hopping integral t and the zero energy level (which is very naturally omitted in all expressions). The spin–orbit interaction will not be considered here, hence, the spin index will be omitted and spin degeneracy is implied. So far, at zero hopping $t=0$ the level has 12-fold degeneracy. The hopping part of the TBA Hamiltonian plays a role of kinetic energy in the Hubbard model. Let us consider non-zero hopping. We will distinguish between hopping along the bond belonging to the triangular and hexagonal bond. The latter seems to be closer to pristine graphene. We will henceforward note two hopping parameters as t_3 and t. Substituting in the Hamiltonian $t_3=0$ we get six non-connected pairs of sites. It is evident that two 6-fold energy levels appear. In the opposite limit of $t=0$ we have four separate triangles with 4- and 8-fold levels.

At the intermediate hopping one has, in general, five levels labeled by tetrahedral symmetry. The degeneracy is 1, 3, 2, 3, 3 from the bottom to top. Three one-electron levels are occupied at half filling. Hence, the $t=0$ case gives the metallic ground state, which becomes a semiconducting one at any small hopping parameter with a gap which is linear in t/t_3 at small t. We note that all these three cases are within a supersymmetry space of the SO(4) problem. Let us consider the maximal 12-dimensional space of the U(12) group (it corresponds to $t=t_3=0$). It can be restricted to the SO(4) subgroup in two ways: [3/2, 1] and [1/2, 5/2] (here we use a standard notation of SO(4) IRR via two angular moments). However, the first representation has only half-integer SO(3) expansions. IRR [1/2, 5/2] has three integer SO(3) representations: $|2\rangle+|3\rangle$, $|1\rangle+|4\rangle$ and $|0\rangle+|5\rangle$ (here bra-vectors give angular momentum IRRs of the SO(3) group). At the same time, the case $t_3=0$ can be presented as the SO(4) state [1/2, 5/2] with $K_z^{(2)}=0$. The reduction of symmetry to the tetrahedral group results in the same set of IRRs for all these integer RRs, which correspond to finite hopping parameters. The case $t=0$ is obtained as a half-integer projection of the SO(4) state. Of course, the one-electron level degeneracy

does not deal with the real spectrum of the system, though classification in terms of hopping kinetic energy may occur as useful.

In the next section we will consider the Frenkel-exciton model as a solution for simplest Hubbard-like Hamiltonian for chargeless electron-hole excitation spectrum.

3. Frenkel-exciton model

We suppose that initially the excitation is localized on a single atom. The electron and a hole on the same site possess a dipole moment, not a charge. Then the resulting excitation is formally an exciton of small radius, a Frenkel-exciton [9]. Let us remind ourselves that in the Frenkel Hamiltonian we preserve, for the chosen two-particle electron-hole state, only four terms – a kinetic energy of an electron, a kinetic energy of a hole, an electron-hole direct Coulomb interaction and an exchange one. The first non-vanishing term in the interaction is a dipole–dipole one:

$$
\begin{aligned}
H_{eff} = &\sum_{1,2} a_1^\dagger a_2 \mathcal{H}_{12}^e - \sum_{1,2} d_1^\dagger d_2 \mathcal{H}_{12}^h \\
&+ \sum_{1,2,3,4} a_1^\dagger d_2^\dagger \mathcal{P}_{12}^* \nabla \frac{1}{|R(1-2)|^3} d_3 a_4 \mathcal{P}_{34}(-\delta_{14}\delta_{32} \\
&+ \delta_{12}\delta_{34}),
\end{aligned} \tag{1}
$$

where the first sum is taken over electron states, the second is taken over hole states; $\mathcal{H}$ represents the kinetic energy operator; $\mathcal{P}$ is a dipole momentum matrix element, taken with the 'bra' and 'ket' vectors of chosen chargeless excitation on site; $|R(1-2)|$ is a distance between sites 1 and 2. Operator nabla appears from multipole expansion of the Coulomb integral as the first dipole term.

The Frenkel-exciton approximation consists of a substitution of exciton operator instead of a pair of electron-hole operators $B_1=d_1a_1$ and in a subsequent linearization of this expression which leads to the following:

$$
\begin{aligned}
H_{FE} = &\sum_1 B_1^\dagger B_1 (\mathcal{H}_{12}^{eh} - W^{(dir)}) + \sum_{1,2} B_1^\dagger B_2 \mathcal{P}_i(1)\mathcal{P}_j^*(2) \\
&\times \frac{\delta_{i,j}|R(1) - R(2)|^2 - 3(R(1)_i - R(2)_i)(R(1)_j - R(2)_j)}{|R(1) - R(2)|^5},
\end{aligned} \tag{2}
$$

where a kinetic energy part and a direct Coulomb interaction are collected into the first diagonal part of the Hamiltonian. The off-diagonal part is given by exchange dipole–dipole interaction between sites 1 and 2.

So far the Frenkel-exciton model is stated based on two parameters $\mathscr{E}=(\mathcal{H}_{12}^{eh} - W^{(dir)})$ and $\mathcal{P}^2/b^3$, where $b \approx 1.44$ Å is the distance between nearest-neighbors in C_{60}. The self energy $\mathscr{E}$ simply shifts the energy zero level. Hence, we will drop it below when it will not cause misunderstanding.

4. Group-theoretical expansion

The harmonic analysis over the group lattice reduces any site-defined problem to ten secular equations for C_{60} (and four for C_{12}). Because of this there are ten IRRs of dimensionality 1, 3, 3, 4, 5 in Y_h (and 4 in T: 1, 1, 1, 3). The remaining task is to write an explicit form of the given Hamiltonian and solve it for all IRRs. We will present below the general consideration but will derive all expressions for the C_{60} lattice, which is more complicated.

The Frenkel Hamiltonian for the Y_h lattice of the C_{60} cluster is given by the sum of the dipole–dipole interaction terms over all 60×60 states:

$$\sum_{g \in Y, f \in \mathscr{F}} P_i^\dagger(g) \frac{1}{|R(f)|^3} \hat{\tau}_{ij}(g, fg) P_j(fg). \qquad (3)$$

Here summation over $g \in Y$ is a summation over 60 sites of the C_{60} cluster surface, while summation over f is restricted by the model over a limited number of neighbors given by some set $\mathscr{F} \subset Y$ (or over a full unlimited set Y, then we will address it as a Hamiltonian of full dipole–dipole interaction). The dipole–dipole interaction between the nearest-neighbors will be considered in the last part of the paper. Then the subset $\mathscr{F} \subset Y$ is given by three fixed elements for each site g (cf. Ref. [2]). $P^\dagger$ and P are independent variables in the secondary quantization representation. In our old notation $P^\dagger(g_1) = \mathscr{P}B_1^\dagger$. We single out a distance dependent factor $1/|R(f)|^3$ from the dipole–dipole interaction and collect an angle dependent remaining part, τ, as it is easily seen from Eq. (2). This τ is actually the traceless second-order tensor which is well-known from classical electric-multipole theory.

The operator $P_i^\dagger(g)$ creates an electron-hole pair on site **g**. Here a vector **g** is directed from the center of the cluster, chosen as the global co-ordinate origin, to the site g. We will use below also the local co-ordinate system (LCS), which will be connected with each site $g \in Y$. The LCS on each site is directed so that any local axes in point $|g\rangle$ goes to $|fg\rangle$ LCS after the proper rotation $f \in Y$.

The operator $P_i^\dagger(g) = a_i^\dagger(g) d^\dagger(g)$ carries an index i connected with the spinor of the electron-hole excitation state.

Let us give now the Frenkel-exciton Hamiltonian in the simplest notation:

$$\sum_{\alpha \in \{RGR \times T_1\}} \sum_{m,n;m_1,n_1 \in \alpha} P_{i,\alpha,m,n}^\dagger$$

$$\times \frac{\tau_{ij}^{(LCS)}(f)}{|R(f)|^3} D_{m_1 k}^{(\alpha)}(f) P_{j,\alpha,m_1,n_1} \delta_{n,n_1} \delta_{m,k}. \qquad (4)$$

where the summation is taken over all indices m, n

belonging to the IRR α containing the direct product of the vector representation T_1 and the full dynamical group of the TB Hamiltonian symmetry, given by the regular representation of Y group (RGR). $P_{i,\alpha,m,n}^\dagger$ is the creation operator of the electron-hole excitation of α IRR (more details are in Ref. [5]). For the sake of clarity we suppose here all bonds to be equal to b, the energy scale is given by the dipole matrix element $\mathscr{P}^2/b^3$. $D_{m_1 k}^{(\alpha)}(f)$, as usual, is the rotation matrix for IRR α. The dipole–dipole interaction in LCS reads as:

$$\tau_{ij}^{(LCS)}(f) = D_{ij}^{(T_1)}(f)$$
$$- 3(D_{i3}^{(T_1)}(e) - D_{i3}^{(T_1)}(f))(D_{3j}^{(T_1)}(f^{-1})$$
$$- D_{3j}^{(T_1)}(e)). \qquad (5)$$

It is seen that the Frenkel-exciton Hamiltonian is reduced to secular equations for each IRR of the following form:

$$\tau_{ij}^{(LCS)}(f) \otimes D_{m_1 k}^{(\alpha)}(f \qquad (6)$$

This is a blockmatrix of dimensionality $3 \times d$, where $d = [\alpha]$ is the dimensionality of an IRR vector. For a non-degenerate mode $\alpha = A$, it is a 3-row matrix. The resulting three full symmetry A-modes have different energies, depending on the local symmetry of a mode. There are five triply degenerate dipole-active modes in the Frenkel-exciton Hamiltonian as it follows from group theory. Our model allows to obtain analytically the mode frequencies as well as the wavevectors for C_{60} and more easily for C_{12} clusters.

5. Summary

We present a Frenkel-exciton model for the calculation of the optical response of carbon nanoclusters of high symmetry. The model Hamiltonian includes a dipole–dipole interaction between electron-hole chargeless excitations. The nearest-neighbor approximation has been used for simplicity of derivation, though the generalization for full dipole interaction is straightforward. The group expansion of the problem is described. The analytical expression will be presented elsewhere for mode frequency and the corresponding wavevectors as a solution of the secular equation of much smaller dimensionality than before expansion. This approach seems to be promising for different lattice problems defined for graphite-like cluster surfaces.

Acknowledgements

This work was partially supported by program 'Fullerenes and Atomic Clusters' project no. 98062, RFBR grant no. 96-02-17926 and grant no. 1-001 of the Russian Program 'Physics of solid state nanostructures'.

References

[1] Auerbach A, et al. Phys Rev Lett 1994;72:2931.
[2] Friedberg R, et al. Phys Rev 1992;B46:14150.
[3] Samuel S. Int Mod Phys B 1993;7(22):3877–97.
[4] Chin TA, Quiang HF. Chem Phys Lett 1995;245:561–5.
[5] Kharlapenko SF, Rotkin SV. The 193th Meeting of the Electrochemical Society; Fullerenes: chemistry, physics and new directions XI. San Diego, CA, May 3–8, 1998.
[6] Rotkin VV. PhD. thesis. St. Petersburg, Russia, 1997.
[7] Rasetti M, Zecchina R. Physica 1993;A 199:539–70.
[8] Chin TA, et al. Chem Phys Lett 1994;227:579–87.
[9] Frenkel JI. Phys Rev 1931;37:17.

Pergamon

Carbon 37 (1999) 851–854

CARBON

Evaluation of various processes for $C_{60}Fe$ production

H. Lange[a,*], P. Byszewski[b,c], E. Kowalska[b], J.Radomska[b], A. Huczko[a],
Z. Kucharski[d]

[a]*Department of Chemistry, Warsaw University, L.Pasteura 1, 02-093 Warsaw, Poland*
[b]*Institute of Vacuum Technology, Długa 44/50, 00-241 Warsaw, Poland*
[c]*Institute of Physics PAS, al.Lotników 32/46, 02-668 Warsaw, Poland*
[d]*Institute of Atomic Energy, 05-400 Świerk, Poland*

Received 16 June 1998; accepted 3 October 1998

Abstract

Iron fullerides were prepared by chemical methods and in the electric arc. The reactions were performed in toluene or benzene solution using aluminium chloride or nitric/acetic acid as activating agents, while ferrocene was used as a source of iron. Iron: carbon plasma was investigated by means of the emission spectroscopy in order to determine the plasma temperature, C_2, Fe and Fe^+ concentrations. The samples were evaluated by mass spectrometry, thermal analyses, X-ray diffraction and by Mössbauer spectroscopy which revealed two doublets ascribed to the iron +2 and +3 ionization states, depending on thermal treatment of the samples. © 1999 Elsevier Science Ltd. All rights reserved.

Keywords: A. Fullerene; C. Mass spectrometry; X-ray diffraction; Mössbauer Spectroscopy

1. Introduction

Fullerides doped with metals with partially occupied 3d-orbitals might exhibit magnetic properties [1] arising from localized magnetic moments, and electric conductivity due to delocalized 4s electrons transferred from the metal to fullerene unoccupied orbitals. It has already been proved experimentally, that the binding energy of iron to C_{60} is so high that $C_{60}Fe$ complexes can be detected by mass spectroscopy in the gas phase and produced in the arc [2–4]. For this reason the geometry, electronic structure and stability of the complexes were evaluated using semi-empirical quantum chemistry model ZINDO1 [5]. The calculations indicated that several possible structures for the $C_{60}Fe$ complex are stable. The optimization procedure converged only if Fe was initially placed above a hexagon or pentagon outside the cage and in the center or close to a hexagon inside the cage. The calculated binding energy was of the order of 900 kJ/mol and depended on the geometry of the complex. Because of the charge shift from C_{60} towards Fe, the complexes exhibit large dipole moments. One might therefore expect that the complexes

should be insoluble in organic solvents such as toluene or benzene and not extractable from the soot.

Here we present our preliminary results for preparation of C_{60} bounded iron by chemical methods and in the electric arc.

2. Experimental results

2.1. Chemical methods.

The preparation of iron containing fullerides by wet chemical reactions were performed using ferrocene (Fc) as a source of iron. They were formed by the reaction of (i) nitrofullerenes-$C_{60}(NO_2)_x$ with Fc, (ii) C_{60} with $Fc^+NO_3^-$ and (iii) from C_{60} with Fc in the presence of $AlCl_3$ as a catalyst. The procedure and results are summed up in Table 1.

1. The nitration of C_{60} was carried out in toluene solution to which a mixture of nitric acid and acetic acid was added (C_{60}/HNO_3 ratio equal to 1/10) with subsequent heating to 95°C for 15 h. The ferrocene dissolved in

*Corresponding author. Fax: +48-22-822-5996.
E-mail address: lanhub@chem.uw.edu.pl (H. Lange)

0008-6223/99/$ – see front matter © 1999 Elsevier Science Ltd. All rights reserved.
PII: S0008-6223(98)00284-X

Table 1
Chemical methods and parameters characterizing the products

	R/C	C_{60}:Fc	heated to 350°C				heated to 500°C			
			Fe/60C	structure	QS	IS	Fe/60C	structure	QS	IS
1	HNO_3	1:2	1.5	a.	1.22	0.44	1.5÷2	a.	1.22	0.44
					2.41	0.53				
2	HNO_3	1:2	2	fcc+a			2÷3	fcc, 1%		
3	$AlCl_3$	1:4	0.7÷1	fcc, 0.5%	2.42	0.53	0.7÷1	fcc, 0.5%	2.42	0.53
									1.22	0.44
4	$AlCl_3$	1:4	1.5÷2	fcc, 1%			1÷1.25	fcc, 0.5%		
	$AlCl_3$	1:10	2	fcc+a.	2.40	0.53	2.5	fcc, 2%		

$1/(C_{60}(NO_2)_x + Fc)$ in toluene:acetic acid at 100°C, 2 h; $2/(C_{60} + Fc^+NO_3^-)$ in toluene:acetic acid at 70°C, 2 h; $3/(C_{60} + Fc)$ in benzene/$AlCl_3$ at 70°C, 4 h; $4/(C_{60} + Fc)$ in toluene/$AlCl_3$, at 15°C, 16 h.
R/C=reagent/catalyst; a=amorphous; (%)-fcc lattice constant increase relatively to pure C_{60}; QS-Mössbauer quadrupole splitting in mm/s (78 K); IS-Mössbauer isomer shift in mm/s (78 K).

toluene was added to the mixture. The amorphous precipitated product, insoluble in standard organic solvents, was used for further analyses. The DSC and TG measurements performed on dried samples exhibited an exothermic peak at 150÷250°C and a mass loss due to solvent removal extending up to 380°C. In the X-ray diffraction on samples annealed in vacuum to 380°C, a strong reflection appears corresponding to diffuse scattering by objects of the size of 1 nm, without traces of iron grains.

2. Iron binding in the ferrocenium ions is weaker than in Fc. Therefore the solution of Fc in toluene:acetic acid:nitric acid, containing $Fc^+NO_3^-$ ferrocenium ions, was added to C_{60} toluene solution. The proportion of active components $HNO_3/Fc/C_{60}$, chosen to minimize polymerization of fullerenes, was equal to 4/2/1. The reactants were heated at 60°C for 1 h. The products still exhibited weak exothermic peaks due to the polymerization of $C_5H_5^-$ and solvent molecules induced by NO_3^- groups. Mass spectroscopy, with electron impact ionization, of fresh samples revealed low mass range peaks which might be ascribed to fragmentation of $C_{10}H_{10}$ thus suggesting that polymerization of cyclopentadienes (cp) occurred already in the toluene solution. X-ray diffraction pattern of the powder consisted of several reflections showing the beginning of crystallization of the product, though the pattern was insufficient to identify the structure. To remove adducts and polymers the samples had to be annealed above 400°C.

3. Boiling of Fc in C_{60} toluene or benzene [6] solution in the presence of $AlCl_3$ may lead to substitution of one cp by a solvent molecule, C_{60} or to formation of a C_{60}Fc adduct. Thus, the pretreated (at 100°C for 3 h) Fc toluene solution was heated with C_{60} toluene solution at 150°C for 16 h in the presence of $AlCl_3$. The ions corresponding to the mass of $C_{60}C_5H_5Fe$ were revealed by LSIMS in the dried product in quantities much exceeding contamination by C_{70}. The samples annealed

above 450°C still contained iron at a concentration of $1Fe/(3÷5)C_{60}$ (depending on the ratio between starting reactants, Table 1), as estimated by X-ray fluorescence. In the samples prepared in benzene solution, distinct endothermic peak was detected by DSC and mass loss by TG at 450°C. The effect occurred close to the decomposition temperature of Fc. Therefore one can suppose that bonding in C_{60}:Fc derivative is of the same strength as in ferrocene. The result suggests that benzene is more suitable for this reaction than toluene.

In the Mössbauer spectroscopy measurements either one or two doublets were detected with QS=2.4mm/s ascribed to Fe^{+2} and 1.22 mm/s ascribed to Fe^{+3}. There were no features in the Mössbauer spectrum, which could be attributed to iron or iron carbide clusters.

2.2. Electric arc method and plasma diagnostics

The plasma experiments were performed in the arc reactor described in details earlier [7] at helium pressure equal to 13.3 kPa and at the arc current 75 A using graphite electrodes of 6 mm diameter. The anode was drilled and packed with iron wires with Fe content equal to 1.3, 4.1, and 5.5 at%. The plasma diagnostics was carried out using the C_2 Swan's band and iron lines. The rotational temperature and concentration of C_2 (a $^3\Pi_u$, $v=0$) radicals, ionization temperature and concentration of atomic and ionic iron were determined.

The presence of iron lines and carbon bands during the discharge indicated that the vaporization of iron wires and carbon electrode took place simultaneously. The C_2 content was determined from the pronounced self-absorption of the transition $d^3\Pi g \rightarrow a^3\Pi_u$, 0–0 (516.52 nm) using the method which allowed to evaluate the rotational temperature and the column density of C_2. The concentration of iron in the gas phase and ionization temperature of iron was calculated from the Saha equation and emission coefficients of iron atomic and ionic lines FeI(516.749 nm)

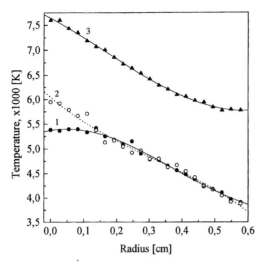

Fig. 1. Radial temperature distributions in the arc. 1 and 2-rotational temperature of C_2 for C and C/Fe plasma, respectively, 3-ionization temperature of Fe (Fe content 5.5 at%).

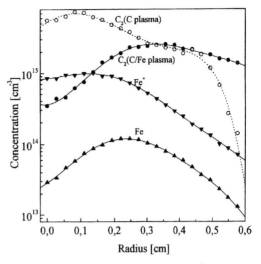

Fig. 2. Radial concentration distributions of C_2(a $^3\Pi_u$, $v=0$), Fe and Fe^+ (Fe content 5.5 at%).

and FeII(516.903 nm), respectively. Because these lines are very close to each other one can expect high accuracy for the determined parameters. The radial temperature distributions are shown in Fig. 1.

It is evident that the ionization temperature exceeds by about 2000 K the rotational temperature, which can be interpreted as the gas temperature because of the efficient energy transfer to the buffer gas. The large temperature difference indicates that the carbon–iron arc plasma is not in a state of local thermodynamic equilibrium. The rotational temperatures of C_2 are nearly the same in pure carbon (line 1) and carbon–iron (line 2) arc. Obviously the translational temperature of iron is close to the gas temperature, because of the high local pressure in the system. The rotational temperature distribution was obtained from non-Abel-inverted spectra. Therefore, the temperature represents the average values. It is mainly related to the arc center, hence the difference at the center between the carbon and carbon–iron plasmas.

The radial concentration distribution of C_2 in the electronic state $a^3\Pi_u$ ($v=0$) and iron species are shown in Fig. 2. The concentration of Fe^+ is much higher than the concentration of Fe and results from the relatively low ionization potential of iron. The low radial gradient of Fe^+ and Fe concentration outside the electrodes suggests that transport is predominantly governed by the gas expansion and thermal diffusion, not by the electric field. The low atomic Fe concentration in the plasma center results from the high ionization degree of iron. The presence of Fe lowers the C_2 content in the arc center, which may probably be connected with the hole in the anode. It is of interest to note that, in the carbon/iron plasma high

concentration of C_2 extends well beyond the arc gap while the temperature remains the same. The effect should influence the process of carbon clustering. Because of comparable densities of iron and C_2 radicals metal:fullerenes are formed in the arc at these experimental conditions.

The toluene and n-hexane extracts from of the soot produced in the arc were analyzed by UV/VIS and X-ray

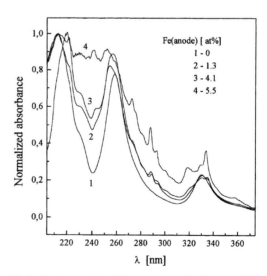

Fig. 3. Absorbance of the fullerene extracts (in n-hexane) of the soot produced with iron-filled electrodes.

fluorescence. The UV–VIS spectra of the extracts are shown in Fig. 3. The numbers at the curves correspond to increasing iron content in the anode: 0, 1.3, 4.1,and 5.5 at%, respectively. The main fullerene absorption features remain, though with higher Fe content in the anode new narrow bands appear at wavelengths where Leach [8] found weak bands denoted by H, F, D, and B. We suppose that presence of iron in clustering carbon gas favors formation of fullerenes with slightly distorted icosahedral symmetry.

Acknowledgements

This work was supported by the funds of the State Committee for Scientific Research Grant No 3 T09A 087 13 (to P.B.) and through the Department of Chemistry, Warsaw University within the project BW-1418/16/98 (to H.L.).

References

[1] Byszewski P, Diduszko R, Baran M. Acta Phys Pol 1994;A85:292.
[2] Roth LM, Huang Y, Schwedler JT, Cassady CJ, Ben-Amotz D, Khar B, Freiser BS. J Am Chem Soc 1991;113:8186.
[3] Huang Y, Freiser BS. J Am Chem Soc 1991;113:9418.
[4] Pradeep T, Kulkarni GU, Vasanthacharya NY, Guru Row TN, Rao CNR. Ind J Chem 1992;31A&B:F17.
[5] Bacon AD, Zerner MC. Theor Chim Acta 1979;53:21.
[6] Byszewski P, Kowalska E, Radomska J, Kucharski Z, Kochkanjan R, Diduszko R, Huczko A, Lange H, Chabanenko V. IWEPNM molecular nanostructures. Kirchberg: Austria, 1998:26.
[7] Lange H, Baranowski P, Huczko A, Byszewski P. Rev Sci Instrum 1997;68(10):3723.
[8] Leach S, Vervloet M, Despres A, Breheret E, Hare JP, Dennis TJ, Kroto HW, Taylor R, Walton DRM. Chem. Phys. 1992;160:451.

Pergamon

Carbon 37 (1999) 855–858

CARBON

Fabrication of aluminum–carbon nanotube composites and their electrical properties

C.L. Xu, B.Q. Wei*, R.Z. Ma, J. Liang, X.K. Ma, D.H. Wu

Department of Mechanical Engineering, Tsinghua University, Beijing 100084, China

Received 16 June 1998; accepted 3 October 1998

Abstract

Aluminum–carbon nanotube (CNT) composites were fabricated by hot-pressing the respective powders. The microstructural characteristics and the distribution of carbon nanotubes in the aluminum matrix were investigated. The electrical resistivities of the composites were measured from room temperature down to 4.2 K. The electrical resistivity at room temperature increases slightly with increasing volume fraction of the carbon nanotubes in the aluminum. From room temperature to ~80 K all the composites exhibit the typical metallic decrease of the electrical resistivity upon lowering the temperature. At about 80 K their resistivities abruptly drop by more than 90%; at lower temperatures the resistivity does not show any fluctuation. © 1999 Elsevier Science Ltd. All rights reserved.

Keywords: A. Carbon nanotubes; Carbon composites; D. Electrical properties

1. Introduction

Carbon nanotubes can be thought of as one or more graphite sheets which have been wrapped up into a seamless cylinder. Since their discovery in 1991 by Iijima [1], the peculiar electronic properties of these structures have received much attention. Their electronic behavior has been predicted [2–4] to depend sensitively on the tubes' diameter and chirality. The electronic properties of multi-walled and single-walled nanotubes have been probed [5–9] experimentally, and show a good correspondence with the theoretical prediction.

In this paper the fabrication of the Al–carbon nanotube composites, the investigation of the microstructure, and the measurement of the electrical resistivity of the composites are reported.

2. Experimental details

The carbon nanotubes used were produced in our laboratory by the catalytic decomposition of the propylene.

The material obtained contains mainly pure carbon nanotubes. Transmission electron microscopy revealed that the tubules have an average diameter of ~30 nm and lengths ranging from hundreds of nanometers to micrometers. Fig. 1 shows a typical TEM microgragh of the used carbon nanotube material. Aluminum powder with an average grain size of ~200 mesh (per square inch) and purity >99.5% were used.

1 wt.%, 4 wt.% and 10 wt.% carbon nanotubes and 35 g aluminum powder were homogeneously mixed by hand grinding for more than 30 min. respectively. The mixtures were hot-pressed at 793 K under a pressure of 25 MPa for more than 30 min. The hot-pressed samples with a size of $\phi40$ mm×6 mm were spark-cut into pieces for various measurements.

Light optical microscopy (LOM), transmission electron microscopy (TEM, JEM-200CX), scanning electron microscopy (SEM, JSM-6301F) and X-ray diffraction (XRD) were used to examine the structure of carbon nanotubes, the distribution of carbon nanotubes and other phases in the composites. The dimensions of samples for measuring the electrical resistivity were $10\times5\times0.5$ mm^3. In order to eliminate the effect of the stray potential and contact resistance, the electrical resistivity at room and at low temperature was measured by the four point method in the Department of Applied Physics, Tsinghua University and

---

*Corresponding author. Tel: +86 10 62782413; fax: +86 10 62787182.

E-mail address: weibq@tsinghua.edu.cn (B.Q. Wei)

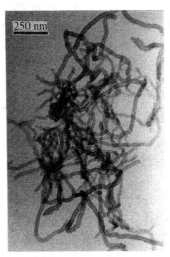

Fig. 1. TEM micrograph of used carbon nanotube raw material

in the National Key Laboratory for Superconductivity in the Institute of Physics, Academia Sinica.

3. Results and discussion

3.1. Microstructure of the composites

Optical microscopy revealed that the carbon nanotubes are located mainly at the aluminum grain boundaries. The hot-pressing processing resulted in a dense microstructure with low porosity. Figure 2a shows a single carbon nanotube embedded in the Al matrix. Also some agglomerates of carbon nanotubes can be found at the grain boundaries of the aluminum matrix.

Table 1
Results of energy dispersive X-ray spectroscopy in the SEM for the sample of Al–4 wt.%CNT (at.%)

measuring points	Al	C	Al:C
1	49.28	50.72	~1:1
2	50.84	49.16	~1:1
3	51.55	48.45	~1:1
4	29.29	70.71	~1:2
5	33.89	66.11	~1:2
6	32.40	67.60	~1:2
7	34.84	65.16	~1:2

Further investigation of the microstructure revealed that also some Al–carbide phases were formed (Fig. 2b). These Al–C phases were found to have an atomic ratio of Al and C atoms of either Al:C=1:1 or 1:2 (Table 1) by energy-dispersive X-ray spectroscopy (EDS) in the SEM. The carbide phases have sizes from tens to hundreds nanometers. In Fig. 2 many carbon nanotubes are found in the close vicinity of AlC or AlC_2 phases, which may partly be formed due to the reaction of amorphous carbon from the surface of nanotubes with the Al matrix.

3.2. Electrical properties of the composites at room temperature

The results of the electrical resistivity measurements of Al, Al–1wt.%, Al–4 wt.%, Al–10 wt.% carbon nanotubes at room temperature are listed in Table 2. There is a slightly increase in electrical resistivity with the addition of carbon nanotubes. Theoretical calculations predicted that the carbon nanotubes have the electrical conductivity of a metal or a semiconductor [2–4]. Experimental measurements of the electrical resistance of individual carbon nanotubes showed an electrical resistance of the order of 10^{-6} and 10^{-4} Ωcm [5,6]. The carbon nanotubes have, in

 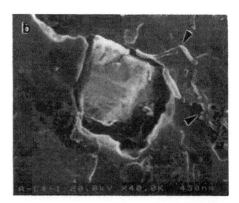

Fig. 2. a. a single carbon nanotube in the Al–1 wt.%CNT sample, b. SEM micrograph of an Al–C phase and carbon nanotubes in close vicinity of it.

Table 2
Electrical resistivity of thin pieces of the samples at room temperature

Sample	Resistivity ($\mu\Omega$cm)
Al	3.4
Al–1 wt.%CNT	4.9
Al–4 wt.%CNT	6.6
Al–10 wt.%CNT	5.5

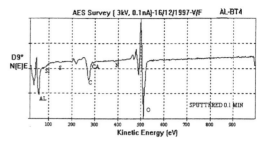

Fig. 4. AES survey results.

any case, a lower electrical conductivity than aluminum. In addition, if the carbon nanotubes are agglomerated at the Al grain boundaries, they can form a kind of grain boundary phase which will increase the scattering of the charge carrier at grain boundary, hence reducing the conductivity. The carbide phases, which are also poorly conducting, and the porosity, which is larger than pure Al, can also influence the electrical resistivity of the composites.

3.3. Electrical properties of the composites at low temperature

The results of the electrical resistivity measurements at low temperature (Fig. 3) are interesting. From room temperature on, the electrical resistivities decreased linearly as the temperature decreased, which is typical metallic behavior. At about 80 K there is an abrupt drop (>90%) in the electrical resistivity of the Al–carbon nanotube composites. Below this transition temperature the resistivity remains at the level of "zero resistance" down to the helium temperature (~4.2 K). Such a phenomenon resembles a superconducting transition.

Hitherto there has been no explanation for these observations. Elemental analysis of the sample surface by Auger electron spectroscopy (AES) showed that there are additional elements on the sample surface mainly oxygen (Fig. 4). However, X-ray diffraction has showed that the bulk material mainly consists of Al and carbon nanotubes (Fig. 5), which indicated that the other phases are minor constituents of the composites. The observed lattice parameter of Al is very close to the reference value of 0.40491 nm. Further experiments on the magnetic properties of the composites are necessary in order to explain the abrupt drop of electrical resistivity at so high a temperature.

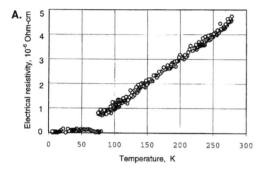

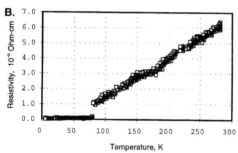

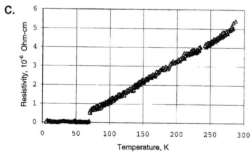

Fig. 3. Dependence of electrical resistivity in Al–carbon nanotube composites on the temperature, a. Al–1 wt.% carbon nanotubes, b. Al–4 wt.% carbon nanotubes, c. Al–10 wt.% carbon nanotubes.

4. Summary

The Al–carbon nanotube composites were successfully fabricated by the hot-pressing process of the respective powders. Al–Carbide phases with composition AlC and AlC_2 were found in the microstructure. The electrical resistivity at room temperature increases slightly with increasing volume fraction of the carbon nanotubes in aluminum. It is very interesting to find an abrupt drop in

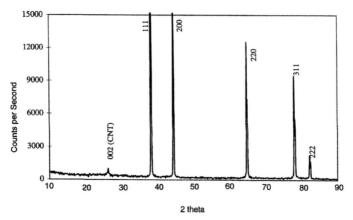

Fig. 5. X-ray diffraction pattern obtained from Al–4 wt.% carbon nanotube sample.

electrical resistivity at about 80 K. Further work is necessary to explain the mechanism of this transition.

Acknowledgements

One of the authors (B.-Q. Wei) wishes to thank the University Stuttgart (Prof. E. Arzt) and the Max Planck Institut für Metallforschung (Prof. M. Ruehle) for support of his research work in Germany. The critical and useful discussions of the manuscript by Dr. Ph. Kohler-Redlich is also acknowledged. This research work was support by National Nature Science Foundation of China under the Grant No. 59501012.

References

[1] Iijima S. Nature 1991;354:56.
[2] Saito R, Fujita M, Dresselhaus G, Dresselhaus MS. Appl Phys Lett 1992;60:22043.
[3] Mintmire JW, Dunlap BI, White CT. Phys Rev Lett 1992;68:631.
[4] Hamada N, Sawada SI, Oshiyama A. Phys Rev Lett 1992;68:1579.
[5] Ebbesen TW, Lezec HJ, Hiura H, Bennett JW, Ghaemi HF, Thio T. Nature 1996;382:54.
[6] Dai HJ, Wong EW, Lieber CM. Science 1996;272:523.
[7] Zhang Z, Lieber CM. Appl Phys Lett 1993;62:2792.
[8] Tans et al SJ. Nature 1997;386:474.
[9] Ge M, Sattler K. Science 1993;260:515.

Pergamon

Carbon 37 (1999) 859–864

CARBON

An investigation of the thermal profiles induced by energetic carbon molecules on a graphite surface

M. Kerford*, R.P. Webb

SCRIBA, School of Electronic Engineering, Information Technology and Mathematics, University of Surrey, Guildford, Surrey GU2 5XH, UK

Received 16 June 1998; accepted 3 October 1998

Abstract

Molecular Dynamics simulation are used to investigate the velocity distributions of a graphite lattice after being struck by carbon molecules. A temperature profile can be inferred from this velocity distribution and the "cooling" down time of the ensuing thermal spike has been investigated. A range of molecule shapes and sizes for different impact energies is investigated. The kinetic energy from the impacts is seen to spread across the surface much faster than into the material in line with the properties of the thermal diffusivity of graphite. A rapid phonon transport mechanism is seen to propagate out from the impact site. The velocity of the wave is found to be independent of the molecule size, shape and energy but the amplitude and start time is dependent on all of these parameters. © 1999 Elsevier Science Ltd. All rights reserved.

Keywords: A. Graphite; D. Phonons; Thermal diffusivity

1. Introduction

The interaction of energetic clusters with solid surfaces is a subject of growing interest due to the increasing use of cluster beams in ion implantation [1] and in ion beam analysis [2]. There has been a great deal of work in this area in recent years, both experimentally [3] and with the use of computer simulation tools [4]. Clusters also provide a method increasing the energy density of the deposition at the surface region of the target way beyond that which is possible by single ion interactions. This makes it possible to produce high energy density, non-linear cascades with ion-target combinations that do not normally experience these.

The impact of a fast moving particle with a solid surface causes the transfer of the particle's kinetic energy to the atoms of the target material. This transfer of energy can result in changes in the bonding configuration of the target material. In graphite targets this can lead to amorphisation as well as promoting changes from sp^2 to sp^3 bonding. Models of the changes made to graphitic targets by impacting particles depend very heavily upon the mechanisms of this energy transfer and the resulting energy propagation through the target. The rate at which the delivered energy is removed from the impact site is important in determining the probability of rearrangement of the area. If the particles in the region remain in motion for only a short time then displacements caused by the initial impact may be "frozen in" to the lattice. If they remain in motion for a long time, however, the lattice may have time to repair itself.

A number of papers have considered this in the past [5,6] but have normally only looked at the effect of the cooling of a thermal spike caused by dumping all of the impacting energy uniformly spread over an ill-defined volume. The study here will use Molecular Dynamics computer simulations to simulate the full collision cascade and observe the subsequent dissipation of the deposited energy.

2. The simulation model

The Molecular Dynamics simulation model used in this study has been described in detail in other publications [7,8] and so only a brief description will be given here. The program uses the Brenner many-body potentials [9] to model the C–C and C–H interactions. For the dynamic systems used here these potentials are known to give both stable graphite and diamond lattices as well as a good

*Corresponding author.

description of the multimer properties of carbon and the correct mechanical properties of graphite and diamond.

The target used in this study is a graphite lattice of dimension 100 Å by 100 Å, consisting of 13 atomic layers. This contains approximately 50 000 atoms. The initial temperature of the lattice is 0 K. Free boundaries are used in conjunction with a large lattice to avoid reflection of energy back from the sides of the simulated micro crystallite. This limits the total time that can be simulated as the energy in the system is not bled away from the micro crystal. However, the main conclusions of this work occur with in a safe time period of the simulation, before much of the energy has reached the sides of the simulated volume.

The simulation solves the equations of motion simultaneously for each particle and stores information on position, velocity and energy states at different time steps. From the kinetic energies of the particles it is possible to calculate a mean square velocity of the simulated target region. From this one could be tempted to infer a mean temperature for the region. This has been done successfully in the past for metal cluster on metal substrates [10,11]. In the first of these, Betz et al. [10], showed that a small region of the target arrives at thermal dynamic equilibrium within the first few 100 fs after the impact and a the velocity distributions of the particles within the region fit a Maxwell-Boltzmann curve. In the second, Colla et al. [11], used a method of gliding averages [12] to describe local temperature variations even when thermal dynamic equilibrium had not been reached for the complete region. As we are interested in the propagation of energy through the target volume we will adopt the latter scheme for this study. Fig. 1 shows the velocity distributions for the whole simulated target volume at different time after the impact of a 4 keV C_{60} molecule. It is clear that the target does not reach a constant temperature until several picoseconds into the simulation. In fact the velocity distributions show that quite quickly the energy of the impacting molecule spreads out but there is a substantial range of temperatures in the simulated region even at quite long times. We, therefore, follow Colla et al. [11] and take a local volume around each particle in the simulation (a 5 Å radius was chosen, although increasing this has little effect on the results presented here). The mean square velocity of the particles inside this volume is then calculated and hence a local "temperature" can be inferred. This enables us to visualise the energy spreading through the target as a function of time.

3. Results and discussion

Simulations of a variety of projectiles have been run, including C_{60}, C_{100} and C_2H_4 for different initial impact energies. Each simulation has been analysed according to the aforementioned methods.

Figure 2 shows a typical impact of a C_{60} molecule on a graphite surface. The example shown has an initial energy of 4 keV viewed at 120 fs after the impact with the surface. As the molecule is large and spherical there is little difference in the propagation of the energy for impacts at different positions on the surface or different orientations of the molecule [13,14]. In this figure each particle is shaded according to the mean square velocity of those particles with in a 5 Å radius. The colour key ranges from white – indicating a "temperature" of more than 750 K through to grey at 0 K. It can be seen that from the figure that the impact creates a small crater and a central core around the crater of particles with high velocities – "hot".

Fig. 2. "Temperature" profile of graphite surface 50 fs after impact of 4 keV C_{60} molecule. Grey scale represents "temperature". White represents >750 K.

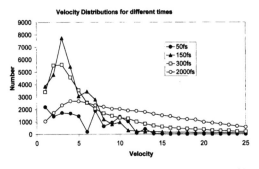

Fig. 1. Velocity Distributions for a 4 keV C_{60} impact on graphite for different times after impact.

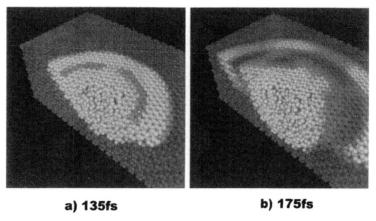

a) 135fs   b) 175fs

Fig. 3. Same plots as Fig. 2 for different times after the impact. (a) 100 fs, (b) 150 fs. Note the break away of wave from "hot" core.

In the example shown the core reaches a maximum "temperature" after only 20 fs of approximately 260 000 K. In this case the number of particles is so small that really all we have is a region of fast moving atoms and the temperature concept is unphysical. However, this region rapidly "cools" and expands to less than 5000 K after 100 fs representing a cooling rate of approximately 2×10^{15} K/s, which agrees with the cooling rate obtained by Marks [6].

As the simulation progresses, a distinct band of increased "temperature" can be seen emanating from the impact site, transferring energy away from the central core. This is shown in Fig. 3a–b. It can be seen that this band of hotter particles spreads out across the surface faster than the central core, which remains hot, gradually spreading its energy out on a longer time scale.

Detailed investigation using computer animation techniques [15] shows that this hotter region propagates as a planar transverse wave. Particles move away from the impact site colliding with atoms around them. The former rebound back to their original sites and the latter continue the propagation. In this way a small amount of energy is transported across the plane of atoms. In contrast, at longer times, an acoustic wave is seen to propagate out from the site as a longitudinal wave, rippling the surface as it travels.

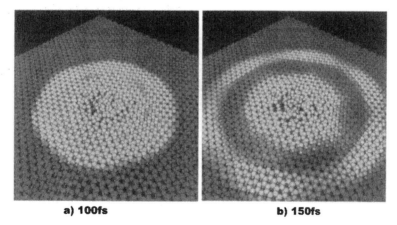

a) 100fs   b) 150fs

Fig. 4. Same plots as Figs. 2 and 3, but sliced through the middle to show the effects for deeper layers. (a) 135 fs, (b) 175 fs. Note the wave propagates along the layers not across the layers.

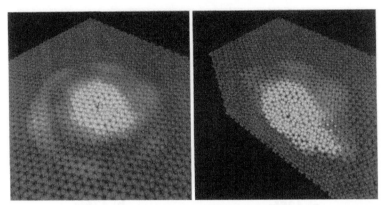

Fig. 5. "Temperature" profiles for 1 keV C_2H_4 impact on graphite after 120 fs. Note the wave appears to be developing sub surface in this case.

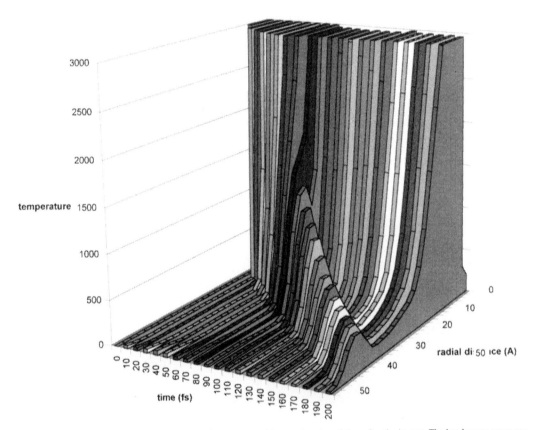

Fig. 6. Radial "temperature" profiles for a 4 keV C_{60} impact on graphite as a function of time after the impact. The break away wave can clearly be seen.

Figure 4a and b show the target sliced through to give an internal view of what is happening. This enables us to see that the wave spreads along the surface of the target, but not down through the layers. This can be explained by the differing thermal diffusivity of graphite in plane and between the layers [16].

The wave can be clearly seen in the simulation of C_2H_4 although it is much weaker than in the case of C_{60}. As the energy per atom increases, the energy is deposited further into the target. This causes the wave to spread through the sub-surface layers. Due to the weak inter-layer coupling of the graphite system, this does not show on the surface – see Fig. 5.

In all the cases simulated on graphite the "temperature" profiles have a radial appearance – this is **not** the case for a diamond target [17], however. In Fig. 6 the surface "temperature" as a function of radial distance from the impact site for different times after the impact. This particular set is for a 4 keV impact of a C_{60} molecule on a graphite surface. By measuring the progress of the wave with time a wave velocity can be found. Fig. 7 shows the progression of this wave for different initial particle energies and different molecules and sizes. In each case the wave velocity is the same, 2×10^4 ms^{-1} for each of the projectiles. The start time of the wave depends upon the velocity of the impacting molecule. This is partly due to the fact that the simulation starts at time zero with the molecule a fixed distance above the surface, therefore the slower the molecule the later it will be before it impacts the surface.

4. Conclusions

There is evidence of a fast thermal wave propagating from the centre of a hot spike region after energetic molecular bombardment of a graphite surface. This is faster than the acoustic wave seen previously in these simulations [15]. Indeed the acoustic wave can be seen as a physical travelling wave across the surface. The acoustic wave is more pronounced at lower energies of about <1 keV for C_{60}. This faster wave can be seen to be propagating as a longitudinal wave rather than the transverse acoustic wave.

The cooling rates of the central core of the spike are in agreement with other work. This allows plenty of time for melting of the impacted region, but much less time for reordering. The length of time that the core stays hot for can be tailored to a certain extent by changes to the size of the molecule and initial energy of the particle impact.

As the energy of the molecule is increased the surface wave can be seen to become sub-surface.

References

[1] Goto K, Makuo J, Takiuchi D, Sugii T, Yamada I. In: Duggan J.L., Morgan I.L., editors. CP392, applications of accelerators in research and industry. New York: AIP, 1997:937.

[2] see for example: Harris RD, Van Stipdonk MJ, Schweikert EA. Int J Mass Spectroscopy and Ion Proc 1998;174:1163.

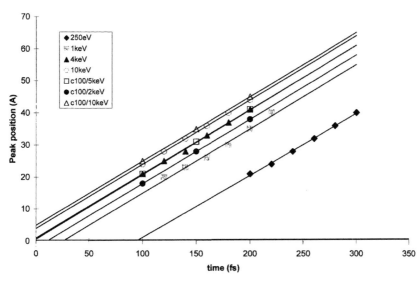

Fig. 7. Progression of the wave as a function of time for different molecules and different initial energies. Note the velocity of the wave is the same only the start times are different – see text for discussion.

[3] Brauchle G, Richard-Scneider S, Illig D, Rockenberger J, Beck RD, Kappes M. Appl Phys Lett 1995;67:52.

[4] see for example recent: Webb R, editor. Computer simulation of radiation effects in solids (COSIRES) proceedings. Radiation Effects and Defects in Solids, Vol. 142 1997

[5] Frauenheim T, Blaudeck P, Stephen U, Jungnickel G. Phys Rev 1993;B48:4823.

[6] Marks NA. Phys Rev 1997;B56(5):2441.

[7] Harrison Jr. DE. Critical Reviews in Solid State Material Science 1988;14:S1.

[8] Smith R, Harrison Jr. DE, Garrison BJ. Phys Rev 1989;B40:93.

[9] Brenner DW. Phys Rev 1992;B46:1948.

[10] Betz G, Husinsky Nu W. Nucl Instr Methods 1997;B122:311.

[11] Colla ThJ, Urbassek HM. Radiation Effects and Defects in Solids 1997;142:439.

[12] Cleveland CL, Landman U. Science 1992;257:355.

[13] Webb RP, Kerford M, Kappes M, Brauchle G. Nucl Instr Methods 1997;B122:318.

[14] Webb R, Smith R, Chakarov I, Beardmore K. Nucl Instr Methods 1996;B112:99.

[15] Webb RP, Smith R, Dawnkaski E, Garrison BJ, Winograd N. Int Video J Eng 1993;3:63.

[16] Handbook of chemistry and physics. 56th ed. CRC Press, Cleveland, Ohio, 1975:E12.

[17] Kerford M, Webb R. (in preparation).

Pergamon

Carbon 37 (1999) 865–869

CARBON

Polycrystalline diamond formation by post-growth ion bombardment of sputter-deposited amorphous carbon films

P. Patsalas[a], S. Logothetidis[a,*], P. Douka[a], M. Gioti[a], G. Stergioudis[a],
Ph. Komninou[a], G. Nouet[b], Th. Karakostas[a]

[a]*Department of Physics, Aristotle University of Thessaloniki, 54006 Thessaloniki, Greece*
[b]*LERMAT, UPRESA-CNRS 6004, 6 Boulevard du Marechal Juin, 14050 Caen Cedex, France*

Received 16 June 1998; accepted 3 October 1998

Abstract

Post-growth Ar^+ ion beam bombardment (IBB) of amorphous carbon (a-C) films on Si, with energies above 1 keV, induces several structural modifications in the films, including the formation of diamond, graphite and SiC grains. X-ray diffraction (in both conventional and grazing incidence geometry) and high resolution electron microscopy were used to study the structure of the as grown films and the phases – with emphasis to diamond – that resulted after IBB. The a-C films morphology and density were also studied by X-ray reflectivity and show an increase in film density upon IBB. © 1999 Elsevier Science Ltd All rights reserved.

Keywords: A. Diamond; B. Plasma sputtering; C. Electron microscopy; X-ray diffraction; D. Microstructure

1. Introduction

Diamond and diamond-like carbon films have received much scientific interest over the last years due to their exceptional properties – such as optical transparency, high hardness, low conductivity, chemical inertness and low friction coefficient – and their potential for technological applications. The diamond films that are deposited by Chemical Vapor Deposition techniques are polycrystalline, but they include a significant amount of hydrogen [1–4]. On the other hand, amorphous carbon films free of hydrogen (a-C), rich in sp^3 bonds that are deposited by Physical Vapor Deposition Techniques (e.g. mass-selected ion beam [5], filtered cathodic vacuum arc (FCVA) [6], laser ablation [7] and bias magnetron sputtering [8,9]) using energetic ions seem to be 'superior' for many applications. However, the lack of diamond grains in a-C films was profound and only recently the formation of diamond grains in a-C films was reported during the ion milling procedure in Transmission Electron Microscopy experiments [10] and in films deposited by FCVA under ion bombardment during deposition [11]. Despite recent

studies of irradiation of a-C:H films which showed graphitization of the film [12,13], in the case of a-C it is predicted theoretically, after thermodynamic considerations, that under irradiation conditions the stable carbon phase could be diamond and not graphite [14]. The sp^3 bonded carbon can form several different crystal structures like the well-known cubic diamond, lonsdaleite (hexagonal diamond-2H) and the hexagonal diamond-6H polytype [15].

Here we report preliminary results of our study on the formation of diamond grains that resulted after a post-growth Ar^+ ion beam bombardment (IBB) of a-C thin films, produced by sputtering, with energy 1.5 keV in conventional IBB experiments or 6 keV during the ion milling procedure for High Resolution Electron Microsopy (HREM) experiments. We present the structure of the as-grown films with regards to the bias voltage (V_b) applied onto the substrate during deposition and the structure of the same films after IBB and we relate the diamond formation with the deposition conditions. We also study the morphology of the films after IBB in terms of their density and surface roughness.

1.1. Experimental

The details of the deposition process have been de-

*Corresponding author. Tel: +3031-998174; fax: +3031-246484.
E-mail address: LOGOT@CCF.AUTH.GR (S. Logothetidis)

scribed elsewhere [8]. Carbon films were deposited by rf magnetron sputtering from a graphite (99.999% purity) target on Si(100) substrates. The carbon films were deposited at various substrate bias voltages ranging from 0 to −200 V at room temperature. Their thickness was 35 (150) nm for those deposited with negative (positive) V_b. All the other parameters were fixed. A phase modulated ellipsometer mounted on the deposition system was used for in-situ Spectroscopic Ellipsometry (SE) measurements in the energy region 1.5–5.5 eV. Following deposition and SE measurements, X-ray Diffraction and X-ray Reflectivity (XRD/XRR) measurements as well as planar view HREM measurements were made. The IBB was conducted in a water-cooled Ultra High Vacuum Chamber, with base pressure better than $5×10^{-10}$ Torr, using a Kaufmann ion source with convergent beam grid geometry. The ion energy was 1.5 keV and the ion beam current 60 mA. The distance between the ion source and specimens was about 25 cm and the corresponding angle was 10°. The working gas (Ar 99.999% purity) pressure was $2.3×10^{-4}$ Torr while the Ar flow rate through the ion source was 5 sccm. After IBB we repeated the XRD/XRR experiments. For the XRD/XRR experiments we used a SIEMENS D-5000 diffractometer equipped with a Goebel mirror [16], a Reflectivity Sample Stage and a Thin Film Attachment (TFA) for grazing incidence asymmetric diffraction. The source was a conventional Cu Kα 2 kW X-ray tube. The Goebel mirror transformed the divergent X-ray beam to parallel. By taking rocking curves of the reflected X-ray beam from the specimen (at angles lower than the critical angle for total external reflection) we calibrated the goniometer with accuracy ~0.001°. The XRD experiments were performed in the form of $\vartheta - 2\vartheta$ locked couple scans (Bragg–Brentano geometry) or in the form of 2ϑ detector scans with fixed angle of incidence between 1–10° (grazing incidence geometry-GIXRD) using the TFA which consists of a series of Soller slits. The generator current and voltage was 40 mA and 40 kV, respectively, the step was 0.02 or 0.04° and the scanspeed was 0.05°/min. The details of the XRR measurements are described elsewhere [17].

Specimens for planar view HREM observations were prepared from samples deposited with $V_b = -20$ V using both chemical etching and ion beam milling.

HREM observations were performed using a TOPCON 002B electron microscope, operating at 200 kV with a point to point resolution of 0.18 nm and $C_s = 0.4$ mm. A more detailed analysis of the HREM observations is described elsewhere [18].

2. Results and discussion

The as-grown sputtered a-C films were found to exhibit very broad XRD reflections, indicative of the amorphous state, at d-spacings that depend on the V_b. In Fig. 1 the

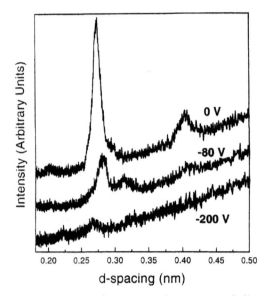

Fig. 1. XRD curves from representative as-grown a-C films deposited at different V_b. The reflections at 0.275 and 0.420 nm are different from those of the (002) graphite and (111) diamond strong reflections.

XRD curves for representative as-grown films deposited at different V_b are shown. It is clear from the figure that the specimens deposited with high negative bias seem to be in amorphous state, while nucleation in the samples deposited with $V_b = 0$ V seems to proceed in turbostratic stacking form [19]. The XRD data are also supported by HREM planar view observations of the same films. HREM observations (Fig. 2) show the existence of nanocrystals that do not belong to the same allotropic form of carbon. Therefore XRD results from these films should contain reflections from various types of crystals. In Fig. 2(a) a HREM image obtained from an a-C film deposited with $V_b = -20$ V in planar view observation prepared by chemical etching is shown. Nanocrystals with d-spacings that cannot be attributed to the same known allotropic carbon forms are shown. For example, a cubic structure is shown in the inset. Based on these observations, in the XRD measurements we will consider that the observed reflections may be attributed to the coexistence of different phases. The XRD results show broad reflections with $d = 0.275$, 0.320 and 0.420 nm that are confirmed by diffraction patterns from HREM. Their intensity depends on the deposition conditions. These reflections are different from the graphite (002) and diamond (111) strong reflections and may correspond to buckminster fullerene[1] or to a

---

[1]JCPDS powder diffraction file card 43-0995.

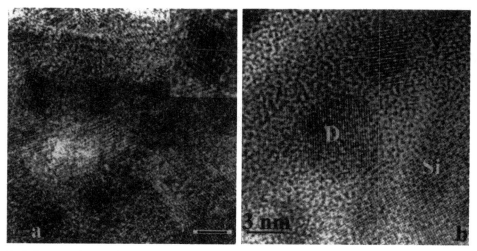

Fig. 2. (a) HREM image of an a-C film deposited with $V_b = -20$ V in planar view observation. Nanocrystals of different structures are formed within an amorphous carbon matrix. A cubic structure is shown in the inset. (b) The same film after ion bombardment. Two nanocrystals, one of SiC and one of diamond are also shown in the amorphous matrix. The coexistence of Si substrate allows the precise determination of the lattice spacings.

cubic form of carbon[2]. A more detailed study of the as-grown films' structure will be presented elsewhere.

In films deposited with $|V_b| < 100$ V, very weak and broad reflections at $d \sim 0.205$ and 0.315 nm, characteristic of sp^3 and sp^2 bonded amorphous material, respectively [20] were also observed. For the films deposited with $V_b = -200$ V the reflections of sp^2 and sp^3 amorphous carbon disappeared and a new, very weak, broad reflection appeared at $d \sim 0.225$ nm. The signal to noise ratio of this peak is very poor. However, this peak was observed, always at the same d-spacing, for several specimens deposited with the same conditions ($V_b = -200$ V) and the reproducibility of the results suggests that it is not noise effect. This change in local order is in agreement with the indications, by SE and XRR studies, of the formation of another carbon phase for films deposited with $|V_b| > 100$ V (ion energy > 130 eV) [21].

After ion bombardment the same films showed well defined reflections, indicative of nanocrystalline material. The broad reflections of the as-grown films have disappeared and new reflections appeared corresponding to different d values, than in the as-grown films. In Fig. 3, a detailed examination of the XRD measurements for d values between 0.20 and 0.27 nm is shown. There exists an intense reflection at about $d \sim 0.235$ nm that corresponds to a reflection of the common hexagonal SiC–2H or –4H.[3]

There are also two other reflections with $d \sim 0.204$ and 0.206 nm and a weak one with $d \sim 0.219$ nm. The strong graphite (002) reflection is absent probably because the graphite grains are textured and are growing along the

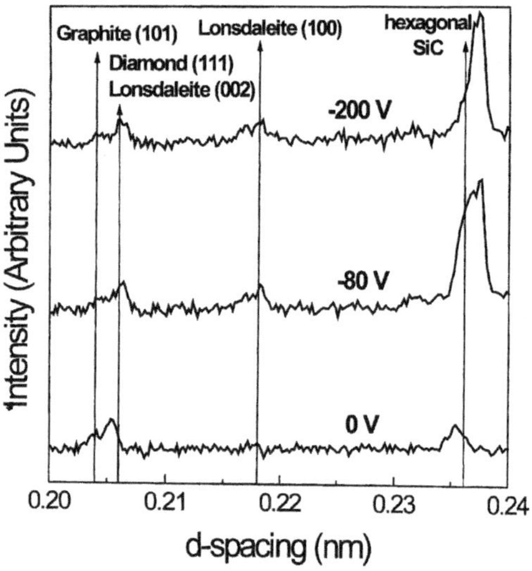

Fig. 3. XRD curves of representative a-C films after IBB. Four reflections are observed that correspond to hexagonal SiC, diamond or lonsdaleite and graphite. The SiC reflection is weaker in the case of the thicker a-C film deposited with $V_b = 0$ V.

---

[2] JCPDS powder diffraction file card 18-0311.
[3] JCPDS powder diffraction file card 43-0995.

(101) direction that corresponds to $d=0.2039$ nm.[4] The reflection at $d=0.206$ nm may correspond either to (111) diamond or to (002) lonsdaleite (hexagonal diamond-2H).[5] Both diamond and lonsdaleite are all sp^3 bonded carbon phases with the same density but different crystal structure [22]. Diamond has the well known cubic diamond structure while lonsdaleite is hexagonal and equivalent to wurtzite. Taking into account the low intensity of the reflection with $d=0.206$ nm it is expected that if this reflection comes from diamond grains then the weak (220) diamond reflection with $d=0.126$ nm (not detected) should be lost in the background. On the other hand, lonsdaleite has strong, characteristic reflections with $d=0.219$ nm (100) and $d=0.206$ nm (002). The intensity of the reflection with $d=0.219$ nm is lower, in all cases, from the intensity of the reflection with $d=0.206$ nm. This suggests that the lonsdaleite grains are textured, with preferred orientation (002), or the reflection with $d=0.206$ nm comes from both cubic diamond and lonsdaleite grains. In GIXRD experiments, these two reflections manifested only for angle of incidence $\vartheta>5°$ (that means that the information is from the deep of the film) suggesting that their origin comes from the a-C/Si interface. The existence of grains that have reflections with $d=0.235$ and 0.206 nm, i.e. hexagonal SiC and cubic diamond or lonsdaleite grains, at the a-C/Si interface is also confirmed by HREM planar view observations obtained from very thin a-C specimens after the ion milling procedure (IBB takes place). In Fig, 2(b) the HREM planar view observation of the a-C film deposited with $V_b=-20$ V after ion milling with a 6 keV Ar$^+$ ion beam is shown. Well defined SiC and diamond nanocrystals, not shown in the as-grown specimens (Fig. 2(a)), that resulted after IBB are observed.

In Fig. 3 we present the XRD curves of representative a-C films after IBB. Four reflections are observed. One ($d=0.235$ nm) corresponds to hexagonal SiC–2H or –4H, one to lonsdaleite ($d=0.219$ nm), one to diamond or lonsdaleite ($d=0.206$ nm) and one to graphite ($d=0.204$ nm). As the bias, during the a-C film growth, becomes more negative, the d value of the diamond (111) or lonsdaleite (002) reflection approaches the value given by PDF data indicating that the grains are less stressed and less deformed.

Information on the density (ρ), thickness, and surface roughness (σ) is deduced by XRR through the dependence of the reflection coefficient on the angle of incidence. The angle of total external reflection, ϑ_c, is related to the refractive index ($n=1-\delta-i\beta$) via the quantity d: $\vartheta_c^2=2\delta =2N_0(e^2/2\pi mc^2)(Z\rho/A)l^2$, where N_0 is Avogadro's number, A is the atomic mass, λ the X-ray wavelength and β is proportional to the linear absorption coefficient. The XRR measurements were analyzed with REFSIM fitting pro-

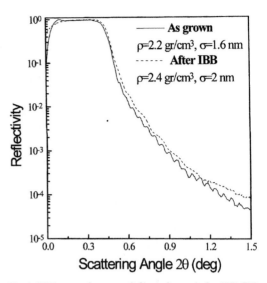

Fig. 4. XRR curves from an a-C film before and after IBB. IBB induces further condensation and surface smoothing in atomic scale.

cedure assuming a single-layer (carbon film/Si-substrate) model that shows very good agreement with experimental results [17]. By measuring the thickness before and after IBB, we calculated the sputtering rate from the a-C films during IBB. The sputtering rate was in the range from 0.8 to 2.0 nm/s depending on V_b during deposition. We also detected a further condensation and surface smoothing, in atomic scale, of the a-C films upon IBB. In Fig. 4 the XRR curves of an a-C film before and after IBB are presented. The density and surface roughness were 2.2 g/cm^3 and 2 nm before IBB and 2.4 g/cm^3 and 1.6 nm after IBB. The sp^3 content of the films before and after IBB was calculated by SE using Bruggeman effective medium theory. This procedure was described [23] and validated using XRR [17,21] elsewhere. SE studies have shown that no considerable change in the sp^3/sp^2 ratio resulted after IBB and only the voids content has changed. Thus, the higher density values may not be due to the formation of diamond grains but due to the collapsing of microvoids in the film, as it was found by SE.

3. Conclusions

The structure of a-C films, developed by magnetron sputtering, was studied by XRD and HREM. The a-C films were found to be mainly amorphous while nuclei of several allotropic forms of carbon were also detected. Post-growth IBB of the above films induced structural modifications including the formation of diamond (either cubic, hexagonal or both), graphite and SiC grains. It was

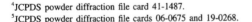

---

[4] JCPDS powder diffraction file card 41-1487.

[5] JCPDS powder diffraction file cards 06-0675 and 19-0268.

found that a-C films rich in sp^3 bonds favored more the formation of diamond grains. No considerable change was found, before and after IBB, in the sp^3 content of the a-C films suggesting that the diamond grains formed by the already existing sp^3 sites in the film and no extra sp^3 content formed by IBB.

Acknowledgements

This work was partially supported by EU EPET II 333 project.

References

[1] Angus JC, Wang Y, Sunkara M. Annu Rev Mater Sci 1998;21:221.

[2] Deshpandey CV, Bunshah RF. J Vac Sci Technol 1989;A 7:2294.

[3] Messier R, Badzian AR, Badzian T, Spear KE, Banchman P, Roy R. Thin Solid Films 1987;153:1.

[4] Spitsyn BV, Bouilov LL, Derjaguin BV. J Cryst Growth 1981;52:219.

[5] Lifshitz Y, Lempert GP, Grossman E. Phys Rev Lett 1994;72:2753.

[6] McKenzie DR, Müller D, Pailthrope BA. Phys Rev Lett 1991;67:773.

[7] Kovarik P, Bourdon EBD, Prince RH. Phys Rev 1993;B 48:12123.

[8] Logothetidis S. Appl Phys Lett 1996;69:158.

[9] Logothetidis S, Gioti M. Mater Sci Eng 1997;B 46:119.

[10] Logothetidis S, Lioutas Ch, Gioti M. Diam Rel Mat 1998;7:449.

[11] Silva SRP, Xu S, Tay BX, Tan HS, Milne WI. Appl Phys Lett 1996;69:491.

[12] Barshilia et al HC. Thin Solid Films 1995;258:123.

[13] Prawer S, Kalish R, Adel M, Richter V. Appl Phys Lett 1986;49:1157.

[14] Zaiser M, Banhart F. Phys Rev Lett 1997;79:3680.

[15] Frenklach M, Kematick R, Huang D, Howard W, Spear KE, Phelps AW, Koba R. J Appl Phys 1989;66:395.

[16] Schuster H, Goebel H. Adv X-ray Anal 1996;39:1.

[17] Logothetidis S, Stergioudis G. Appl Phys Lett 1997;71:2463.

[18] Komninou Ph, Nouet G, Kehagias Th, Logothetidis S, Gioti M, Karakostas Th. Diam Relat Mater 1998: in press.

[19] Bish DL, Post JE. Modern powder diffraction.In: Reviews in mineralogy, vol. 20. Mineralogical Society of America, 1989.

[20] Lioutas CB, Vouroutzis N, Logothetidis S, Boultadakis S. Carbon 1998;36:545.

[21] Logothetidis S, Patsalas P, Gioti M. Phys Rev Lett 1998:submitted.

[22] Bundy FP, Kasper JS. J Chem Phys 1967;46:3437.

[23] Gioti M, Logothetidis S. Diam Rel Mat 1998;7:444.

Pergamon

Carbon 37 (1999) 871–876

CARBON

Electrical behaviour of metal/a-C/Si and metal/CN/Si devices

E. Evangelou[a,*], N. Konofaos[a], S. Logothetidis[b], M. Gioti[b]

[a]*Physics Department, Applied Physics Laboratory, University of Ioannina, 45110 Ioannina, Greece*
[b]*Physics Department, Aristotle University of Thessaloniki, 54006 Thessaloniki, Greece*

Received 16 June 1998; accepted 2 October 1998

Abstract

Electrical characterisation of metal/carbon/Si devices was performed. Amorphous carbon films rich in sp^3 bonds were grown onto n-type Si substrates by RF magnetron sputtering at room temperature. Different deposition conditions were used to create different sp^3 and sp^2 configurations in order to examine their influence on the performance of electronic devices. Suitable metalisation was used to fabricate devices, which were then characterised electrically. Electrical characterisation using I–V, C–G–V and G–ω techniques showed temperature dependent currents through the devices which increase rapidly when forward bias is applied. This behaviour was found to be dependent on the sp^3–sp^2 contents of the films. The devices behaved like metal–insulator–semiconductor diodes with a defect insulator resulting in creating thermally activated currents through the devices. The effect of nitrogen introduced in the growth process to produce carbon nitride films was also examined. Different amounts of nitrogen were used and the same characterisation process has been used for a variety of samples. The films were nearly perfect insulators and the corresponding devices showed a clear MIS behaviour. Thus, the room temperature magnetron sputtering technique produced films, with electronic properties dependent on the C–C bonding configuration. Moreover it is shown that the nitrogenated films made under certain conditions can be used as insulators in devices. © 1999 Elsevier Science Ltd. All rights reserved.

Keywords: A. Amorphous carbon; Carbon films; B. Sputtering; D. Electronic properties

1. Introduction

The application of carbon based materials into microelectronic devices has been of considerable interest in the last few years. Tetrahedral amorphous carbon (ta-C) films have recently shown electronic properties suitable for making heterojunction devices with silicon, as the CVD grown diamond-like films have shown before [1–3]. N-type ta-C films grown by FCVA could also be obtained using nitrogen as a donor dopant [4]. The room temperature (RT) rf-magnetron sputtering technique has been proved very efficient in the production of films rich in sp^3 bonds [5]. Films with graphite-like structure as well as nitrogenated films can also be grown [6,7]. Its main advantage is the cold substrate which is of great importance for the electronic components manufacture since it minimises temperature induced structural modifications to both the silicon substrate and the deposited carbon films. Using suitable masks, this method can be used to build

devices if the electronic behaviour of the produced films can be identified.

The aim of this work is to determine the dependence of the conductive properties of amorphous carbon (a-C) and carbon nitride (CN_x) films on the deposition conditions and the corresponding applications to the electronic devices.

2. Experimental

Both n-doped and undoped carbon films were grown and investigated. The method of growth was the rf-magnetron sputtering at room temperature. The amorphous carbon films were deposited on n-type Si (100) wafers from a graphite target (99.999%) at room temperature (RT). The working gas during deposition was Ar (99.999% purity), its flux controlled by mass flow controllers. The kinetic energy of the Ar^+ ions during etching was controlled by the bias voltage applied at the substrate in order to be as low as possible so that the amorphisation of the Si substrate was minimised. This process was controlled with

---

*Corresponding author.

0008-6223/99/$ – see front matter © 1999 Elsevier Science Ltd. All rights reserved.
PII: S0008-6223(98)00288-7

Table 1
Characteristics of a-C samples

Sample	Applied substrate bias (V)	Thickness Å	sp^3/sp^2	ρ (Ωcm)
#1	+10	2175	1/99	1×10^7
#2	0	1300	5/100	2×10^7
#3	0	980	48/57	1×10^{13}
#4[a]	+10/−20	913	85/31	7×10^{14}

[a] Sample #4 was grown in sequential layers with positive and negative bias.

the real-time Spectroscopic Ellipsometry technique. For this, a phase modulated ellipsometer attached to the deposition chamber at an angle of incidence of 70.4° which allows in-situ and real-time measurements in the energy region of 1.5–5.5 eV was used.

Nitrogenated films were produced by the injection of nitrogen gas inside the chamber during growth. Suitable flow controllers regulated the whole procedure. N_2 was introduced into the chamber at a flow rate in the range of 2 to 30 sccm, thus resulting in various N_2 plasma concentrations. The gas was of 99.999% purity. The distance between the target and the substrate was 65 mm and the discharge power at the graphite target was 100 W. All films were deposited at room temperature [6,7].

Ohmic back contacts of low resistance were formed on the n-Si substrate by aluminium alloying. After careful degreasing of the chips, metal contacts were fabricated on top of the films by evaporative deposition of 1mm diameter Al dots in an oil-free vacuum system using suitable masks.

The in-situ spectroscopic ellipsometry (SE) was used to measure the pseudo-dielectric function of the films. The pseudo-dielectric functions were analysed by means of the Bruggenman Effective Medium Theory (BEMT) in combination of the three phase model (air/composite a-C (CN_x) film/c-Si substrate) and procedures described elsewhere [5,6,8]. In this analysis we assume that the composite films consist of sp^3 [9], sp^2 [9] bonds and voids and we deduced the film thickness and the volume fractions of sp^3, sp^2 sites.

Seven series of samples, four with undoped and three with doped films were prepared and examined. The growth conditions, together with the optically calculated parameters of the samples are summarised in Table 1 for the a-C

films and in Table 2 for the CN_x films. It is important to note here that the sp^3 and sp^2 volume fractions presented in Tables 1 and 2 are not a quantitative measure but they are rather suitable for comparisons between the films. In Table 2 the sum of the sp^3 and sp^2 fractions of the samples #2, 3 and 4 is above 100% due to the fact that the materials are more dense than that used as the reference in the analysis of SE data.

A four-probe station connected to an HP4192A LF impedance analyser and a Keithley 617 pico-ammeter formed the measurement system. A suitable cryostat equipped with temperature controllers and regulators was used for the temperature dependent measurements. More details about the experimental set-up are presented elsewhere [10].

3. Results and discussion

Figure 1 presents the RT I–V characteristics of the devices containing the a-C films. The resistivity of the films was extracted from the corresponding I–V curves and included in Table 1. The second column of that table contains the sp^3/sp^2 C–C volume fractions ratio for each sample.

Sample #1 is highly conductive, resulting in providing I–V curves similar to heterojunction devices. The extreme behaviour of that sample, was further examined by studying the effect of temperature on the I–V characteristics. These data are plotted in Fig. 2 and show a marginal change of current with temperature. Although these curves may indicate the formation of a heterojunction, the corresponding C–V curves at different frequencies show a rather poor MIS structure instead [11]. Therefore, the exact

Table 2
Characteristics of CN_x samples

Sample	Nitrogen concentration the plasma at %	Thickness (Å)	sp3/sp2	ρ (Ωcm)
#5	1.04	1940	43/34	3×10^{14}
#6	9.76	1880	51/32	4×10^{13}
#7	32	1875	51/34	1×10^{13}

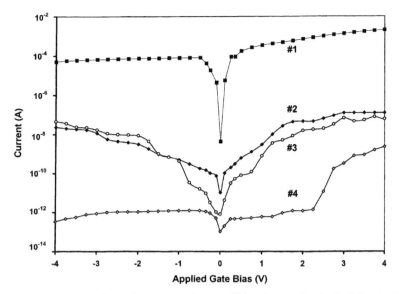

Fig. 1. Room temperature current–voltage characteristics for the a-C devices: ■ Sample #1, ♦ Sample #2, ○ Sample #3, ☆ Sample #4.

conductivity mechanism is of controversial origin and is currently under investigation. Sample #1, a clearly conducting sample, has also the lowest sp^3/sp^2 ratio measured. A gradual temperature increase from 77 K up to RT results to a change of the reverse currents of nearly four

orders of magnitude. At the same time the forward current increases only one order of magnitude. This is a typical behaviour of bulk related leakage currents through the carbon films. Thus, the thermally activated currents must be attributed to the bulk properties of the carbon films

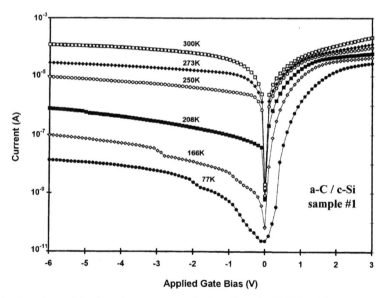

Fig. 2. The current–voltage characteristics for various temperatures for the a-C sample #1. ● T=77 K, ☆ T=166 K, ■ T=208 K, ○ T=250 K, ♦ T=273 K, T=300 K.

which in this case act as a defect insulator, nearly approaching the behaviour of an intrinsic semiconductor [12].

Regarding the rest of the a-C devices, sample #4 can be considered as a good insulator. We should note here that a-C film #4 was grown in sequential layers with alternative positive (+10 V) and negative (−20 V) bias voltage on the substrate during deposition. Not surprisingly, this is also the sample with the highest sp^3/sp^2 ratio. The other two samples exhibit properties in between samples #1 and #4. In order to study the AC behaviour of the devices Capacitance and Conductance versus voltage (C–G–V) characterisation with frequency as a parameter was used in order to examine the AC properties of the devices [8]. This method is very commonly used on MIS (metal–insulator–semiconductor) structures. The results of the C–G–V profiling are used in verifying the insulating behaviour of the material sandwiched between the semiconductor and the metal. In these kind of devices, the top electrodes are commonly known as 'gate electrodes' because of their potential use in MOSFET type devices. The characteristics appear in Fig. 3 for sample #2. They are the typical curves of MIS devices with the carbon films at the insulator position.

Thus, the growth conditions and especially the applied substrate bias, control the C–C bonding configurations [6] and regulate the electronic properties of the produced films. The control over the deposition parameters could provide either good insulators or intrinsic semiconductors. The resistivity of the films increases as the C–C bonding approaches that of diamond. The sp^3-rich a-C films can be used as insulating layers in electronic devices. On the other hand, rich in sp^2 bonds a-C films can be used in Schottky type devices.

Table 2 shows the deposition conditions and the relative results for the CN_x films, together with the nitrogen concentration in the plasma. The corresponding I–V curves taken at room temperature are plotted in Fig. 4. The CN_x films exhibit a behaviour which can be directly related to the amount of nitrogen involved in the deposition process whereas there are not significant variations on the calculated sp^3/sp^2 ratio which are presented in Table 2. All three samples break down at fields of the order of 10^5 V/cm. Below that field the films grown with lower N_2 concentration show negligible current flow in contrast with sample #7 which grown with considerably more nitrogen. In particular, the devices produced are either MIS diodes when a low percentage of nitrogen is introduced in the plasma (samples #5 and #6), or very poorly performing diodes when this percentage is around 30% (sample #7). Sample #7 conducts only when the applied bias is positive, indicating a rectifying junction between the CN_x and Si. However, it is important to note that this characteristic is obtained without any post-thermal annealing of the samples, a procedure quite common when dopants are introduced in materials by other methods in order to activate the carriers [3]. The AC behaviour of the films was examined using C–G–V profiling. The typical characteristics of sample #6 are plotted in Fig. 5. They confirm that the devices behave like MIS diodes.

The question raising is about the top metal contacts if such materials are to be used in Schottky or junction

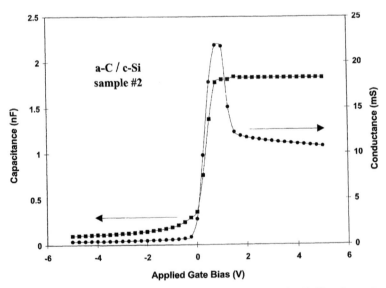

Fig. 3. C-V (■) and G-V (●) curves for the frequency of 1 MHz for the a-C device sample #2. They show a clear MIS behaviour.

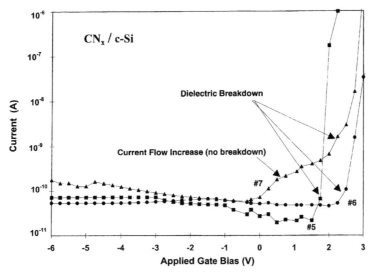

Fig. 4. Room temperature current–voltage characteristics for the a-C devices: ■ Sample #1, ♦ Sample #2, ○ Sample #3, ☆ Sample #4.

devices. The metal contacts need to be examined on whether they are purely ohmic or rectifying regarding to the doping profiles of the CN_x films [3]. This problem is currently pursued, with samples of different nitrogen concentrations and different metals on top [11].

Compared to recently published results on similar materials grown by various techniques (ta-C:H [13], ta-C:B [14], CN_x [15] and polycrystalline films of carbon [16]), our results look compatible. The measured values of resistivities for all these films vary according to the deposition conditions and the dopants. The measured conductivities of such films can be attributed either to the bonding (like in the case of our a-C films) or to the presence of the dopant atoms and their properties [17]. The

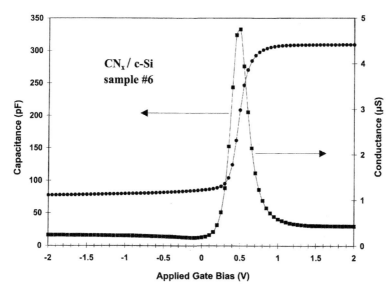

Fig. 5. C-V (■) and G-V (●) curves for the frequency of 1 MHz for the CN_x device sample #6. They show a clear MIS behaviour.

latter explanation seems the most suitable for our doped films as well.

4. Conclusions.

The a-C and the CN_x films grown by rf magnetron sputtering at RT can be used in electronic devices with silicon, behaving either as semiconductors or (in most cases) like insulators. The behaviour is strongly dependent on the ratio of sp^3/sp^2 bonds. On the other hand, doped films, are expected to behave like n-type semiconductors when nitrogen is involved. Our results show that low concentration of nitrogen in the plasma during deposition, actually enhances the insulating behaviour of the films, making MIS diodes instead. When that concentration exceeds a threshold value, CN_x layers become semiconducting (as expected) and the corresponding devices are heterojunction diodes.

The main advantage of the whole procedure is the ability to create films with different conductive properties deposited on silicon only by altering external parameters, such as the applied substrate bias and the nitrogen flux. In this way, carbon films with different conductivities can be grown while the substrate is kept at room temperature. This avoids any alterations of the structure and properties of the formerly deposited films.

References

[1] Kalish R. Appl Surf Sci 1997;117–118:558.
[2] Chhowalla M, Robertson J, Chen CW, Silva SRP, Davis CA, Amaratunga GAJ, Milne WI. J Appl Phys 1997;81:139.
[3] Konofaos N, Thomas CB. J Appl Phys 1997;81:6238.
[4] Clough FJ, Milne WI, Kleinsorge B, Robertson J, Amaratunga GAJ, Roy BN. Electron Lett 1996;32:498.
[5] Logothetidis S. Appl Phys Lett 1996;69:158.
[6] Gioti M, Logothetidis S. Diam Rel Mat 1998;7:444.
[7] Logothetidis S, Lefakis H, Gioti M. Carbon 1998;36:757.
[8] Aspnes DE. Thin Soid Films 1982;89:249.
[9] Logothetidis S, Petalas J, Ves S. J Appl Phys 1996;79:1040.
[10] Logothetidis S, Evangelou E, Konofaos N. J Appl Phys 1997;82:5017.
[11] Konofaos N, Evangelou E, Logothetidis S, Gioti M. Proc. of the XIV Greek Solid State Physics Conference. Ioannina, Greece, 1998.
[12] Sze SM. Physics of semiconductor devices, 2nd Edition. New York: Wiley, 1981.
[13] Conway NMJ, Milne WI, Robertson J. Diam Rel Mater 1998;7:477.
[14] Kleinsorge B, Ilie A, Chhowalla M, Fukarek W, Milne WI, Robertson J. Diam Rel Mater 1998;7:472.
[15] Takada N, Arai K, Nitta S, Nonomura S. Appl Surf Sci 1997;113–114:274.
[16] Boettger E, Bluhm A, Jiang X, Schaefer L, Klages CP. J Appl Phys 1995;77:6332.
[17] Stumm P, Drabold DA, Fedders PA. J Appl Phys 1997;81:1289.

Pergamon

Carbon 37 (1999) 877–880

CARBON

Comparison of fullerene–iron complexes modeling with experimental results

E. Kowalska[a], Z. Kucharski[b,*], P. Byszewski[a,c]

[a]*Institute of Vacuum Technology, ul.Długa 44/50, 00-241 Warsaw, Poland*
[b]*Institute of Atomic Energy, 05-400 Świerk, Poland*
[c]*Institute of Physics PAS, al. Lotników 32/46, 02-668 Warsaw, Poland*

Received 16 June 1998; accepted 3 October 1998

Abstract

We have applied the semiempirical quantum chemistry ZINDO method in order to evaluate the binding energy of fullerenes with Fe and several ferrocene derivatives that might be formed from a ferrocene solution in toluene or benzene and optimized possible FeC_{60} complexes of various structures and spin states. Information on the bonding energy helped us to estimate thermal treatment conditions necessary to purify the samples of the oligomers formed during the reaction of ferrocene with fullerene. The Mössbauer spectroscopy was used to investigate iron state in the FeC_{60} samples. These results are discussed using the charge distribution calculated in the ZINDO model. © 1999 Elsevier Science Ltd. All rights reserved.

Keywords: A. Fullerene; C. Mössbauer spectroscopy; Molecular simulation; Modelling

1. Introduction

Great research interest in the synthesis and characterization of fullerenes derivatives and in transition metal fullerene chemistry [1,2] have led to the discovery of new organometallic compounds. In some of them the metal is directly bound to carbon on the fullerene cage and the ligands of the parent organometallic compound remain as part of the new one.

We search for transition metal: fullerene compounds, free of any hydrocarbon complexes, whose electrical and magnetic properties would be determined by interaction of the metal d-electrons and fullerene π- electrons. The experiments began with iron because we hoped that applying Mössbauer spectroscopy, the state of the metal in the compound could be determined. The fact that FeC_{60} complexes may exist was established by Pradeep et. al [3] who also determined some properties of FeC_{60} in a solid form. Yet a simple synthesis method of the host–guest $C_{60}(ferrocene)_2$ compound [4] has shown that iron and fullerite can form a stable compound thus providing iron containing fullerite for further experimental investigations [5,6]. The main goal of this paper is to discuss Mössbauer spectroscopy results within the terms of semiempirical quantum chemistry models.

2. Experimental

2.1. Mössbauer spectroscopy

The Mössbauer spectra were recorded in a standard transmission geometry using a constant acceleration spectrometer coupled to a 50 mCi $^{57}Co/Rh$ source (Amersham, Buchler). A thin absorber layer was used; typically the samples contained less than 0.1 mg of the ^{57}Fe isotope per 1 cm^2. Isomer shifts are given relative to metallic iron (α-Fe) at room temperature. The Mössbauer cryostat was a helium bath cryostat MD-306 Oxford Instruments coupled to the temperature controller ITC-4 Oxford Instruments. In such systems the temperature stability was better than 0.1 K. The spectra were computed with a least-squares routine using Lorentzian lines.

2.2. Samples preparation

We prepared two different types of sample denoted as A and P. In both cases the reactions were aimed to bind

---

*Corresponding author.

ferrocene (Fc) or its derivative to C_{60} thus preventing its evaporation from the fullerite based solid during annealing. It was expected that the consecutive annealing would lead to evaporation of the residual solvent, decomposition of the iron carrier and evaporation of the ligands. In the method A fullerenes had been nitrated in toluene solution to which a mixture of nitric acid and acetic acid was added at the molar ratio $C_{60}/HNO_3 = 1/10$. The solution was heated to 95° C for 15h then reacted with ferrocene dissolved in toluene ($C_{60}/Fc = 1/2$). The amorphous precipitate annealed in vacuum to various temperatures was used for further experiments. The diffuse X-ray scattering and electron diffraction showed that, the solid annealed in vacuum consisted of the objects of the size of ~1 nm suggesting that fullerenes sustained the treatment.

Samples P were prepared using the method of ferrocene ligand substitution by solvent molecules occurring in the presence of $AlCl_3$ [7,8]. The iron diarene and arene-Fecp molecules are unstable unless they are coordinated with an electron acceptor. The reaction was carried out in the C_{60} and Fc toluene solution ($C_{60}/Fc = 1/20$) therefore $C_{60}Fc$ [9] or C_{60}Fecp complexes could also be formed. The solution was dried and annealed at 350°C in vacuum in order to remove residual solvent and free Fc.

Iron concentration in the final products was determined by an X-ray fluorescence method. The A type sample used for the Mössbauer spectroscopy contained 1.7Fe/60C and the P type 2.5Fe/60C. Further technological details of both methods are described in [10].

3. Results and conclusions

The typical liquid nitrogen temperature Mössbauer spectra of the P and A fullerene/iron compounds are presented in Fig. 1. The very clear quadrupole doublet

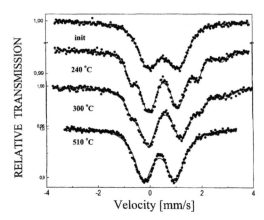

Fig. 2. Mössbauer transmission spectra measured at 78 K of the sample A in initial form heated to 240°C, 300°C and 510°C.

observed in compound P correspond to Mössbauer parameters: QS=2.396 mm/s and IS=0.532 mm/s, at the temperature $T=78$ K. These values are characteristic of the low spin iron Fe^{+2} and are very close to that found in $C_{60}Fe_2$, $C_{60}Fc_2$ [5,6] and pure Fc. Another doublet with the QS approximately half the size: ≈ 1.20 mm/s and IS=0.43 mm/s at $T=78$ K was observed in the sample A, the values usually observed in materials containing Fe^{+3} ions.

The thermal treatment of the samples modifies the spectrum (Fig. 2). Beside the main doublet there appears, in the sample A, a second one with QS~2.4 mm/s whose intensity at first increases with the annealing temperature then the doublet disappeares. If one assumes that the sample contained any Fc derivatives then these changes may be ascribed to the variation of iron ionization states. There could be ferrocenium cations in the fresh sample because of the nitric acid used for the reaction, it returned

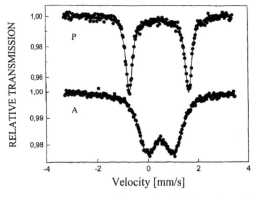

Fig. 1. Mössbauer spectra of the P (top) and A (bottom) sample recorded at liquid nitrogen temperature. The full curves are computed fits.

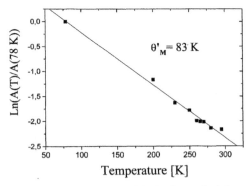

Fig. 3. Temperature dependence of the ln of normalized absorption for P. sample. The full lines are computed fits.

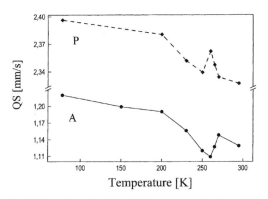

Fig. 4. The quadrupole splitting (QS) temperature dependence in P and A sample.

to neutral Fc's during low temperature annealing and at high temperatures it decomposed leaving iron in the lattice. In the fresh P sample there exists only the ferrocene like

doublet whose intensity diminishes with the annealing temperature while the new one with QS=1.22 mm/s appears with gradually increasing intensity.

As is evident from the experimental data (Fig. 3) the intensity of the Fe^{+2} doublet changes with the temperature. It points to the relatively low Mössbauer lattice temperature (Q_M) i. e. weak iron coupling to the lattice. The Mössbauer lattice temperature can be calculated, using the high temperature limit of the Debye model, from the equitation [11]:

$$\Theta_M = \frac{E_\gamma}{c} \sqrt{-\frac{3}{M_{eff} k d(\ln A)/dT}}$$

where E_g, c, k, M_{eff}, A are the energy of γ radiation, the light velocity, the Boltzman constant, the Mössbauer nucleus atomic effective mass and the normalized absorption, respectively.

The standard fitting procedure applied to the data presented in Fig. 3 give $Q_M = 83$ K in contrast to 130 K in $C_{60}Fc_2$ and 91 K in pure Fc [5,6].

It is interesting to note that both samples exhibited

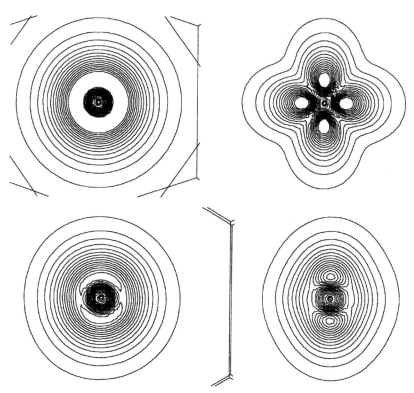

Fig. 5. Front and side projection of charge distribution isosurfaces around Fe in ferrocene (top) and FeC_{60}(h), S=2 complex. (bottom); cp and fullerene fragments show scale of the drawing. Isosurfaces start at 0 step 0.15 to 2.25 and down to 0 in $[ea_0^{-3}]$.

similar QS temperature dependence with large fluctuation around 260 K (Fig. 4). We ascribe such behavior to the fullerene hindered rotation influence on the charge distribution around Fe nuclei.

The interpretation of the Mössbauer experiments are ambiguous because there is no information on sites occupied by iron in the fullerite lattice. In order to find sites favored by iron in the fullerite lattice we have performed a semiempirical quantum calculation of various FeC_{60} complexes using the ZINDO model implemented in the commercially available package. It was possible to optimize complexes of four different conformations, i.e., exohedral Fe opposite the C_{60} hexagon (FeC_{60} h), opposite the pentagon and an endohedral one with Fe in the center of C_{60} and shifted towards the hexagon. Because the charge distribution around the iron nucleus determines the QS it was calculated for the known Fc derivatives as well as for the optimized FeC_{60} complexes. The highest electron density in Fc (of D_{5d} symmetry) forms a toroid around Fe parallel to the ligands (Fig. 5, top). Similar distribution was found in all considered Fc derivatives where QS is close to 2.4 mm/s yet not in C_{60}Fc where the C_{60} size distorts Fc symmetry. The charge distribution in all FeC_{60} complexes but FeC_{60} h, spin S=2, substantially differs from the toroid resembling a octahedrally distorted sphere. The charge distribution in FeC_{60} h, spin S=2 is shown in the bottom part of Fig. 5. The asymmetry of the complex is evidently compensated by one of the Fe 4s electrons not participating in the bonding. Unfortunately we have not succeeded in optimizing any $C_{60}FeC_{60}$ complex, probably because of the unrealistic starting structure. Yet we managed to optimize the $C_{40}H_{10}FeC_{40}H_{10}$ complex where C_{40} was made of a part of C_{60} with hydrogen compensating the broken bonds. Although calculations were started with the linear conformation and the pentagons facing Fe, the energy minimum was reached at the angle $C_{40}H_{10}$–Fe–$C_{40}H_{10}$ equal to 120°. The corresponding charge distribution in the σ_v plane (neglecting internal $C_{40}H_{10}$ structure) resembles that in Fc depicted in the top left part of Fig. 5. These results suggest that the Fc like Mössbauer spectrum may originate from iron at very special low symmetry sites but not from

Fc directly attached to C_{60}. The other doublet with low QS may arise from Fe at the other sites or Fc adducts with the charge distribution resembling octahedrally distorted sphere. Neither the spherical nor the octahedral part contributes to the quadrupole splitting, which remains low. It seems that the iron sites characterized by large QS are metastable and their occupation decreases with the annealing temperature.

Acknowledgements

This work was supported by State Committee for Scientific Research Grant No 7 T08A 016 10 and 3 T09A 087 13.

References

[1] Balch AL, Catalano VJ, Costa DA, Fawcet WR, Federco M, Ginwalla AS, Lee JW, Olmstead MM, Noll BC, Winkler K. J Phys Chem Solids 1997;58:1633.

[2] Balch AL. In: Taylor R, editor. The Chemistry of Fullerenes. River Edge, New Jersey: World Scientific, 1995.

[3] Pradeep T, Kulkarni GU, Vasanthacharya NY, Guru Row TN, Rao CNR. Ind J Chem 1992;31A&B:F17.

[4] Crane JD, Hitchcock PB, Kroto HW, Taylor R, Walton DRM. J Chem Soc Chem Commun 1992;1992:1765.

[5] Byszewski P, Kucharski Z, Suwalski J. In: Kuzmany H, Fink J, Mehring M, Roth S, editors. Progress in fullerene research. New Jersey:World Scientific, 1994:82.

[6] Birkett PR, Kordatos K, Crane JD, Herber RH. J Phys Chem B 1997;1997:8975.

[7] Nesmiejanov AN. In: Nauka, editor. Chemistry of iron, manganese and rhenium σ- and π-complexes. Moscow, 1980 (in Russian).

[8] Pruchnik F. In: PWN, editor. Organometallic chemistry, transition elements. Warsaw, 1991 (in Polish).

[9] Olah GA, Bucsi I, Aniszfeld R, Surya Prakash GK. In: Kroto HW, Fisher JE, Cox DE, editors. The fullerenes. Pergamonn Press, 1993:65.

[10] Lange H, Byszewski P, Kowalska E, Radomska J, Huczko A, Kucharski Z. Carbon 1999;37:851.

[11] Herber R, Maeda Y. Physica 1980;B99:352.

Pergamon

Carbon 37 (1999) I

CARBON

Keywords for Carbon

Authors should select a maximum of five keywords. Each keyword should be accompanied by the capital letter denoting the category from which the keyword has been selected, as shown in the following example.

Keywords: A. Carbon composites; B. Impregnation, Pyrolysis; C. Ultrasonic measurements; D. Mechanical properties.

A. Types of Carbon

activated carbon
amorphous carbon
anthracite
carbon aerogel
carbon beads
carbon black
carbon cloth
carbon clusters
carbon composites
carbon fibers
carbon filaments
carbon films
carbon microbeads
carbon nanotubes
carbon onions
carbon precursor
carbon xerogels
carbon/carbon composites
carbyne
catalyst support
catalytically grown carbon
char
charcoal
coal
coal tar pitch
coke
diamond
diamond-like carbon
doped carbons
electrodes
exfoliated graphite
fullerene
glass-like carbon
graphite
graphite oxide
highly oriented graphite
intercalation compounds
isotropic carbon
isotropic graphite
kish graphite
mesophase
mesophase pitch
natural graphite
needle coke
nuclear graphite
petroleum coke
petroleum pitch
pitch
pyrolytic carbon
resins
semicoke
single crystals
soot
synthetic graphite
tars
vapor grown carbon
whetlerite
whiskers

B. Preparation and Processing

activation
annealing
baking
calcination
carbonization
chemical treatment
chemical vapor deposition
chemical vapor infiltration
coating
coking
combustion
cracking
doping
electrochemical treatment
etching
gasification
graphitization
grinding
halogenation
heat treatment
high pressure
hydrothermal treatment
implantation
impregnation
infiltration
intercalation
laser irradiation
mixing
oxidation
plasma deposition
plasma reactions
plasma sputtering
pyrolysis
pyrolytic deposition
sintering
stabilization
surface treatment
thermosetting

C. Techniques

adsorption
atomic force microscopy (AFM)
BET surface area
chemisorption
chromatography
differential scanning calorimetry (DSC)
DRIFTS
dynamical mechanical thermal analysis
(DMTA)
ESCA
electrochemical analysis
electron diffraction
electron energy loss spectroscopy (EELS)
electron microscopy
electron paramagnetic resonance (EPR)
electron spin resonance
ellipsometry
image analysis
infrared spectroscopy
ion scattering spectroscopy (ISS)
light scattering
mass spectrometry
microcalorimetry
modelling
Mössbauer spectroscopy
molecular simulation
nuclear magnetic resonance (NMR)
neutron scattering
optical microscopy
photoelectron spectroscopy
Raman spectroscopy
rheometry
scanning electron microscopy (SEM)
scanning tunneling microscopy (STM)
spectrophotometry
temperature programmed desorption (TPD)
thermal analysis (DTA and TGA)
thermodynamic simulation
transmission electron microscopy (TEM)
ultrasonic measurements
x-ray diffraction
x-ray photoelectron spectroscopy (XPS)
x-ray scattering
XFAS (EXAFS)

D. Properties and Phenomena

acoustical properties
adsorption properties
aggregation
biocompatibility
carbon yield
catalytic properties
chemical structure
coke yield
crystal structure
crystallite size
defects
dielectric properties
diffusion
elastic properties
electrical (electronic) properties
electrochemical properties
electronic structure
field emission
fracture
frictional properties
functional groups
galvanomagnetic properties
gas storage
heat of adsorption
immersion enthalpy
intercalation reactions
interfacial properties
lattice constant
lattice dynamics
luminescence
magnetic properties
mechanical properties
microporosity
microstructure
Mössbauer effect
optical properties
particle size
phase equilibria
phase transitions
phonons
photoconductivity
porosity
radiation damage
reaction kinetics
reactivity
specific heat
superconductivity
superlattices
surface areas
surface oxygen complexes
surface properties
textures
thermal conductivity
thermal diffusivity
thermal expansion
thermodynamic properties
transport properties
viscoelasticity

Carbon – Instructions to Authors

(An International Journal Sponsored by the American Carbon Society)

A more detailed set of instructions can be found in the journals home page at: **http://www.elsevier.nl/locate/carbon.**

Submission of Papers

Authors are requested to submit their original manuscript and figures with two copies to a regional Associate Editor, if such exists or to one of the Editors-in-Chief: Prof. P.A. Thrower, c/o Elsevier Science, The Boulevard, Langford Lane, Kidlington, Oxford OX5 1GB, UK or Prof. Ljubisa R. Radovic, 205 Hosler Building, Department of Energy and Geo-Environmental Engineering, The Pennsylvania State University, University Park, PA 16802, USA.

Addresses for submission to Regional Associate Editors:
H.P. Boehm, Institut für Anorganische Chemie der Universität München, D-80333 München 2, Meiserstrasse 1, Germany;
T. Burchell, Oak Ridge National Laboratory, Lockheed Martin Energy Research, Inc., PO Box 2008, Oak Ridge, TN 37831-6088, USA;
P. Ehrburger, Institut de Chimie des Surfaces et Interfaces (ICSI), 15 rue Jean Starcky, BP 2488, 68057 Mulhouse, France;
M. Inagaki, Hokkaido University, Faculty of Engineering, Kita-ku, North 13 West 8, Sapporo 060, Japan;
B. McEnaney, School of Materials Science, University of Bath, Bath BA2 7AY, UK;
F. Rodriguez-Reinoso, Department of Inorganic Chemistry, University of Alicante, Apartado 99, Alicante, Spain

In some cases it may be appropriate to submit a manuscript to the Associate Editor whose research interests are related to its subject matter. Letters to the Editor must be sent to Prof. P.A. Thrower. All papers are to be submitted in English. Submission of a paper implies that it has not been published previously, that it is not under consideration for publication elsewhere, and that if accepted it will not be published elsewhere in the same form, in English or in any other language, without the written consent of the publisher.

Types of Contributions

Papers should deal with original research work in the physics, chemistry, and scientific aspects of technology of a class of materials that range from diamond and graphite through chars, semicokes, mesophasic substances, carbons, carbon fibers, and fullerenes. They should communicate new knowledge in adequately finished form and should not be deficient in scientific information (e.g. as to type of material investigated).

Manuscript Preparation

General: Manuscripts must be typewritten, double-spaced with wide margins on one side of white paper. Good quality printouts with a font size of 12 or 10 pt are required. The corresponding author should be identified, include a Fax number and E-mail address. Full postal addresses must be given for all co-authors. Authors should consult a recent issue of the journal for style if possible. An electronic copy of the paper should accompany the final version. The Editors reserve the right to adjust style to certain standards of uniformity. Original manuscripts are discarded one month after publication unless the Publisher is asked to return original material after use.

Abstracts: An Abstract not exceeding 200 words, should be provided for full papers: letters to the editor do not have an abstract. A list of key words should follow the abstract, or, in the case of a letter to the editor, the title.

Text: Follow this order when typing manuscripts: Title, Authors, Affiliations, Abstract, Keywords, Main text, Acknowledgements, Appendix, References, Vitae, Figure Captions and then Tables. The corresponding author should be identified with an asterisk and footnote. All other footnotes (except for table footnotes) should be identified with superscript Arabic numbers. The paper should only contain a detailed technical description of methods used when such methods are new.

Units: The SI system should be used for all scientific and laboratory data.

References: All publications cited in the text should be presented in a list of references following the text of the manuscript. In the text refer to references by a number in square brackets on the line (e.g. Since Peterson [1]), and the full reference should be given in a numerical list at the end of the paper. References should be given in the following form:

1. Palmer HB, Cullis CF. In Walker Jr. PL, editor. Chemistry and Physics of Carbon, Vol. 1, New York: Marcel Dekker, 1965. p. 265–266.
2. Fuertes AB, Marban G, Muniz J. Carbon 1996;34(6):223–230.
3. White JB, Brown FH, Jones C. Extended Abstracts 20th Biennial Conference on Carbon. Santa Barbara, CA, 1991. p. 348.

Illustrations: All illustrations should be provided in camera-ready form, suitable for reproduction (which may include reduction) without retouching. Photographs, charts and diagrams are all to be referred to as 'Figure(s)' and should be numbered consecutively in the order to which they are referred. They should accompany the manuscript, but should not be included within the text. All illustrations should be clearly marked on the back with the figure number and the author's name. All figures are to have a caption. Captions should be supplied on a separate sheet.

Line drawings: Good quality printouts on white paper produced in black ink are required. All lettering, graph lines and points on graphs should be sufficiently large and bold to permit reproduction when the diagram has been reduced to a size suitable for inclusion in the journal. Dye-line prints or photocopies are not suitable for reproduction. Do not use any type of shading on computer-generated illustrations.

Photographs: Photographs should only be included where they are essential. Original photographs must be supplied as they are to be reproduced (e.g. black and white or colour). If necessary, a scale should be marked on the photograph. Please note that photocopies of photographs are not acceptable.

Colour: Authors will be charged for colour at current printing costs.

Tables: Tables should be numbered consecutively and given a suitable caption and each table typed on a separate sheet. Footnotes to tables should be typed below the table and should be referred to by superscript lowercase letters. No vertical rules should be used.

Electronic Submission

Authors should submit an electronic copy of their paper with the final version of the manuscript. The electronic copy should match the hardcopy exactly.

Proofs

Proofs will be sent to the author (first named author if no corresponding author is identified of multi-authored papers) and should be returned within 48 hours of receipt.

Letters to the Editor: Letters to the Editor are published rapidly, and once accepted will be typeset and prepared as camera ready copy for rapid publication, by Prof. P. A. Thrower. For this reason all figures, both line drawings and photographs, must be submitted in a size to fit within a 7.5 cm column width. Figures will not be reduced. There is usually insufficient time for proofs to be sent to authors by mail for correction, and they are often faxed. Please enclose your Fax number with your original submission. If contact with the authors cannot be made, every care is taken to ensure accuracy by having the proof checked by other people. Please understand that this procedure is necessary to expedite publication.

Offprints

Twenty-five offprints will be supplied free of charge. Offprints and copies of the issue can be ordered at a specially reduced rate using the order form sent to the corresponding author after the manuscript has been accepted. Authors requiring extra offprints of a Letter to the Editor should indicate this with their original submission. Orders for reprints will incur a 50% surcharge.

Copyright

All authors must sign the 'Transfer of Copyright' agreement before the article can be published. This transfer agreement enables Elsevier Science Ltd to protect the copyrighted material for the authors, but does not relinquish the author's proprietary rights. The copyright transfer covers the exclusive rights to reproduce and distribute the article, including reprints, photographic reproductions, microfilm or any other reproductions of similar nature and translations. Includes the right to adapt the article for use in conjunction with computer systems and programs, including reproduction or publication in machine-readable form and incorporation in retrieval systems. Authors are responsible for obtaining from the copyright holder permission to reproduce any figures for which copyright exists.

Author Services

For queries relating to the general submission of manuscripts (including electronic text and artwork) and the status of accepted manuscripts, please contact Author Services, Log-in Department, Elsevier Science, The Boulevard, Langford Lane, Kidlington, Oxford OX5 1GB, UK. E-mail: **authors@elsevier.co.uk**, Fax: +44 (0) 1865 843905, Tel: +44 (0) 1865 843900.

Printed and bound by CPI Group (UK) Ltd, Croydon, CR0 4YY

08/05/2025

01864849-0003